CREATIVE &
MODERN
OFFICE Ⅲ

Edited by HI-DESIGN PUBLISHING

Dalian University of Technology Press

CREATIVE & MODERN OFFICE III

Copyright ©2014 by
Dalian University of Technology Press

Published by
Dalian University of Technology Press

Address: Section B, Sic-Tech Building, No.80 Software
Park Road, Ganjingzi District, Dalian, China
Tel: +86 411 84709043
Fax: +86 411 84709246
E-mail: yuanbinbooks@gmail.com
URL: http://www.dutp.cn

All rights reserved. No part of this publication may be
reproduced or transmitted in any form or by any means,
electronic or mechanical, including photocopying,
recording or any other information storage and retrieval
system, without prior permission in writing by the publisher.

ISBN 978-7-5611-9144-6

PREFACE 01

It was only thirty years ago when a personal computer Apple Lisa appeared. For the first time it used a graphical user interface and a mouse pointing device for individual business users. Since then we can say offices began to change fast as information technology has given a power kick-start to various branches of human activity, which resulted in dozens previously unthinkable professions. All those professions have one trend in common – they all are creative, focused on the intelligent, virtual product. These changes embodied in a new type of workspaces – offices for creative people. This kind of workspace expects lack of hard dress code and standard working day, the ability to switch from work to rest and vice versa. Employers had to revise their views on how workspace should look, which areas in the office are exactly required, which corporate values it should represent. Another challenge was to attract highly skilled employees, for which, as it turned out not only the location of the office was an important factor in particular, but its usability, ergonomics, recreational opportunities, visual appeal as well.

We are one of the few companies in our country known primarily due to our innovative approaches in workspace construction. Our clients are mainly IT companies that come to us for "an unusual bright office". Every time it's a complex, but exciting challenge – because it often happens that we have to use unusual methods and materials to achieve a certain visual effect or a non-trivial design. Merely in last two years, while working on office projects we had a chance to pick up pickling solutions for copper sheets, worked with a magnetic coating, all kinds of acoustic materials, developed sails and hundred-meter lamps. We had to create submarine shaped flower pots, equipped offices with bars, covered walls and ceilings with carpet and designed polymer structures.

Our offices are filled with bright colors. Varnished concrete, original furniture and all kinds of complex geometric structures can be found there. As architects we should keep in mind that such a powerful "visual range" is a kind of responsibility. This responsibility extends far beyond the customer and his wishes – because there are hundreds or even thousands of employees who will stay in this workspace the most part of the day. Therefore, the major priority for us is to create humane conditions for employees and to take care about a variety of opportunities for relaxation and activity changing. Maybe this is a kind of self-limitation, perhaps even against our own creative impulses and ambitions.

The architecture is still based on the three Vitruvian's principles as it was two thousand years ago – durability, utility and beauty. Beauty, no wonder, is in the last place, so while creating a modern office, one should not forget the fact – beauty is important, but its role is secondary to functionality and comfort. I would generally expand Vitruvian triad in reference to the office design, and add to it the concept of "humanity" – as one of the most important criteria of the creative workspace.

Peter Zaytsev (Right)
Architect and Co-owner of za bor architects

PREFACE 02

Thanks in large part to technological advances, the modern office is an increasingly connected one; collaboration now drives productivity, as opposed to solitary work. As a result, the landscape of the office itself is changing to be more open and collaborative.

Ideas flow freely when barriers are removed. It's no surprise then that private offices are less important, while benching systems and open office layouts become more popular every year. With the removal of these barriers, we're also seeing the modern office become more democratized. Whereas in previous eras, windows with views and natural light were reserved for executives with elusive corner offices, the contemporary workspace now is more likely to open up these views to everyone. It's not unusual to find CEO working out in the open office and collaborating directly with employees.

The global workforce as a whole is also more mobile than ever, and requires an equally flexible workspace. There is a greater need for spaces that can accommodate multiple functions – a work cafe, for example, that can also house all-hands meetings during the day and movie nights in the evening, or even, as we've seen in one case, a reception desk that doubles as a bar counter during happy hour. While a variety of design trends are at play in the modern office, from homey and rustic to sleek and modern, the overall focus is on making the office a flexible and welcoming space, one where users feel both comfortable and productive.

New workspaces also set forth the challenge to create more efficient, green, and earth-friendly environments. Companies are expressing a greater interest in efficient fixtures and recycled or multifunctional materials, which reduce energy costs, keep materials out of landfills, and reduce stress on the environment.

Whether the company is technology, finance, or retail based, new workplaces reveal an increasing focus on innovation – how to work in faster, smarter, and more exciting ways. The modern office not only encourages collaboration, but also gives users space to grow and develop better ways of working in the future.

Melissa Wallin
CEO of Design Blitz
Seth Hanley
Creative Director of Design Blitz

CONTENTS

8
GOOGLE OFFICE

22
ONE WORKPLACE HEADQUARTERS

34
EBAY GITTIGIDIYOR

46
OGILVY & MATHER, JAKARTA

56
TENCENT DAZU OFFICE

64
YANDEX SAINT PETERSBURG OFFICE II

76
SKYPE'S NORTH AMERICAN HEADQUARTERS

86
CASTROL OFFICE

96
ROPEMAKER

102
360 HQ OFFICE

110
ENGINE INNOVATION LABS

116
GOOGLE CSG SUPER HQ

126
UNILEVER OFFICE

136
JWT

144
RED BULL

Google Dublin – A thriving new campus boosting the spirit of innovation

Google Ireland opens the doors to its thriving new campus: Four buildings located in the heart of Dublin's historic docklands district! With over 47,000 m^2 of unique office space, the campus represents an amazing workplace for Google's ever growing sales, marketing, finance and engineering teams, coming from more than 65 countries and speaking over 45 languages.

The tallest of the four campus buildings is the newly constructed 14-storey building "Google Docks" which is also the tallest commercial building in Dublin. Two other buildings – "Gasworks House" and "Gordon House" – have been home for Google already and were completely refitted. The fourth building, "One Grand Canal" – fondly referrred to as "1GC" – was newly fitted out.

For the Masterplan the architects had to find a smart solution for the nearly impossible – to create a stimulating and interactive campus within a bustling environment in the midst of the inner city. Apart from innovative office spaces, the Masterplan required the successful organization of a multitude of additional functions, such as 5 restaurants, 42 micro kitchens and communication hubs, game rooms, fitness center, pool, wellness areas, conference, learning & development center, tech stops, over 400 informal and formal meeting rooms and phone booths, etc.

All these additional functions are part of the holistic work philosophy of Google, encouraging a balanced, healthy work environment and enabling as much interaction and communication between the Googlers as possible. In fact, as empirical researches and studies show, interaction and communication are crucial for creativity and innovation.

In order to achieve these important goals, the locations of all functions were very carefully balanced between the different floors and buildings. In addition, a bridge is planned to be built to connect "Google Docks", "Gasworks House" and "Gordon House" together, encouraging an easy flow between buildings and people.

LOCATION Santa Clara, California, USA

ONE WORKPLACE HEADQUARTERS

DESIGNER
Design Blitz

AREA
3,252 m²

PHOTOGRAPHER
Bruce Damonte

When One Workplace decided to move its headquarters and showroom, it had grand ambitions for its new location. Interestingly enough, the company succeeded in its visionary goals by embracing the past and renovating an older industrial warehouse and office that had fallen into disrepair. One Workplace collaborated with Design Blitz, a young architecture firm with a track record of creating innovative offices for successful tech companies, to create a bleeding edge, world class workplace. Design Blitz's design for the new facade and landscape improvements expand One Workplace's space into a multifunctional indoor-outdoor environment that transcends both office and showroom norms.

During initial conversations with the client, Design Blitz began mapping both the client and user experiences through the space. The mapped experiences demonstrated that both customers and users would be sent out into the space to experiences a series of carefully planned touch points and then brought back to their starting point – much like the path taken by a boomerang. The boomerang further manifested itself in the creation of the two-storey stacked "boomerang" mezzanine in the center of the open office space. The elevated conference room and observation platform allow members of the One Workplace team to quickly survey the floor and show customers how a variety of systems solutions can intermix to create a unified, flexible and layered approach to workplace layout.

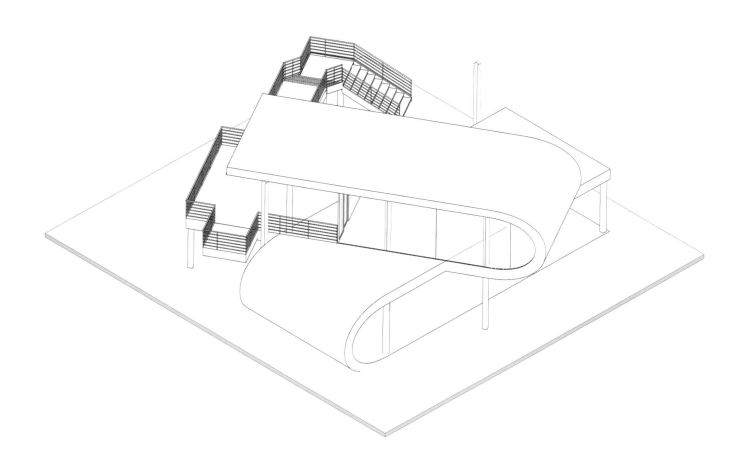

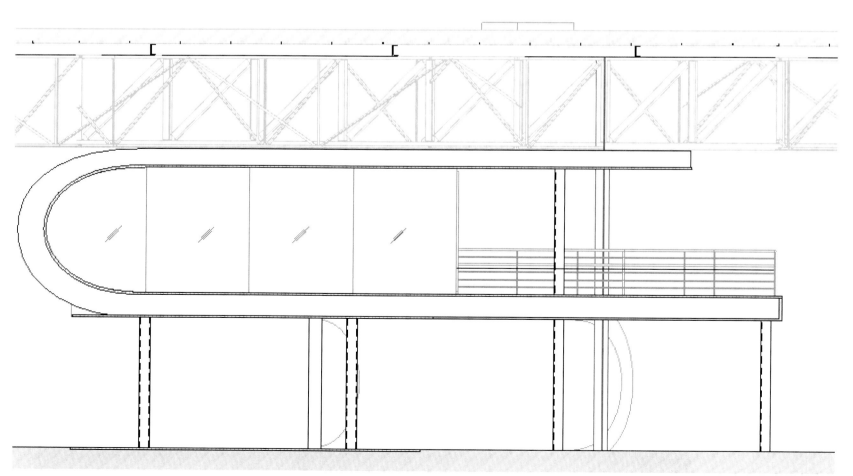

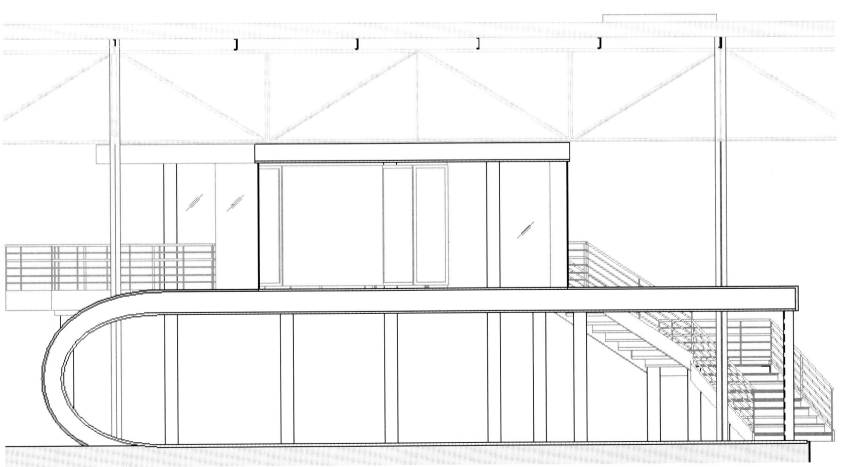

LOCATION Istanbul, Turkey

EBAY GITTIGIDIYOR

DESIGNER
O.S.O Architecture & Interior Design

AREA
2,000 m²

PHOTOGRAPHY
Gürkan Akay

The new office of "eBay – GittiGidiyor" which is one of the most important players of e-trade in global and local markets is located in My Office Building in Istanbul Atasehir. In line with the global trend of change in today's modern offices, eBay Istanbul office is planned as an "open office". Accordingly, the inside office areas can be described by 4 functions: the entrance hall & social facilities, open office, meeting rooms and technical & service areas.

The entrance hall that will build the initial perception of visitors for the office is designed as an impressive & inviting place. The desired "inviting sense" is emphasized by the natural wood work at the floor and ceiling and by the reception desk positioned at the back. The wooden pergola representing the entrance hall invites people to walk through the reception.

The "social place" positioned behind the entrance hall is the only place visitors allowed to see, besides the meeting rooms. So, the place welcoming the visitors and used for celebrations of office staff in-house also includes various services like: cafe-bar, library, on-line music, TV, projection and play station games. A terrace is related with this place which can also be used as a smoking area.

The open space working area holds 164 staff on 1,100 m^2 with a maximum of 225 seats. The ceiling design of the open space – which is a challenge to overcome the typical acoustical problems encountered in such areas, reflects an irrational order of lighting instruments and acoustical panels.

There are 12 meeting rooms with different sizes, 7 of which are used externally. The hi-tech infrastructures of the rooms are export from UK. Acoustical panels are used in ceiling design of these rooms where international meetings are held.

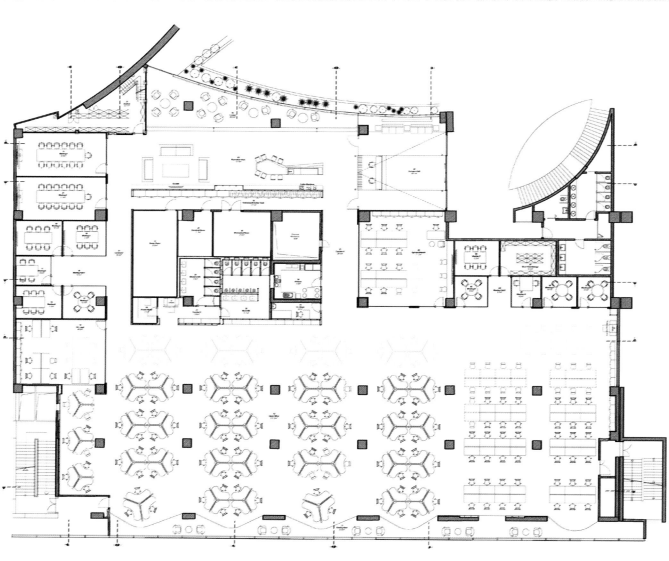

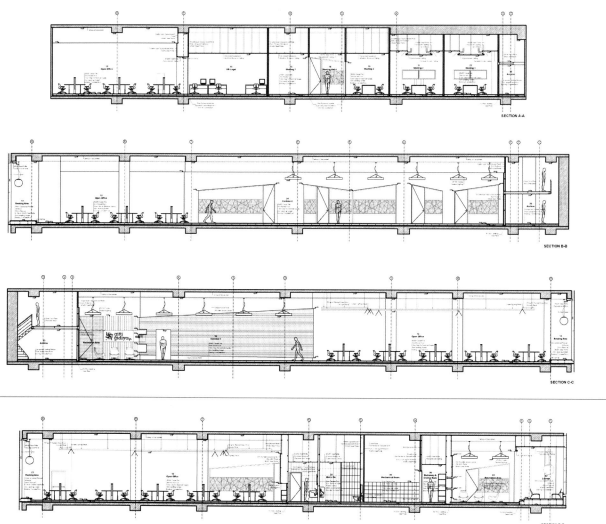

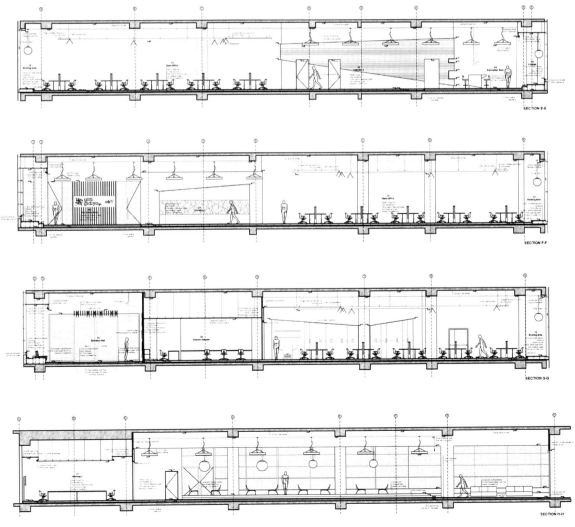

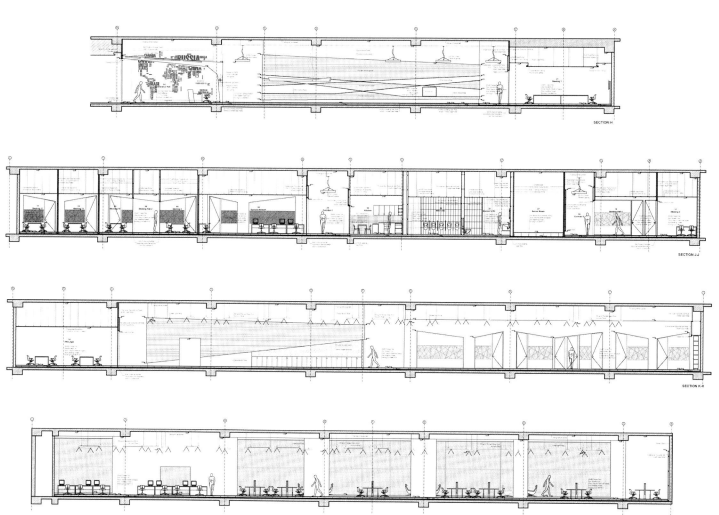

LOCATION Jakarta, Indonesia

OGILVY & MATHER, JAKARTA

DESIGNER
M Moser Associates

AREA
2,603 m²

PHOTOGRAPHER
Jack Shea

The need for this workplace initially arose from Ogilvy & Mather's desire to consolidate its multiple Jakarta offices into a single location. As well as cutting property expenses, bringing all staff together under one roof would improve teamwork and creativity, and bring coherence to Ogilvy's brand image in the city.

Collaborating closely with Ogilvy, the M Moser design team examined the firm's operational dynamics in depth to refine the project brief. Key requirements emerged for a mixture of open and enclosed spaces, and for a multipurpose café/meeting area.

Ogilvy's desire for a contemporary Indonesian aesthetic inspired the "Bali spa" atmosphere of the 11th floor reception. Sitting atop a low timber platform, the glass-topped desk is supported by eight solid timber discs. "Silk cocoon" fixtures hang from the ceiling above, with another trio of fixtures in the background marking the top of a new staircase to level 10.

Flanking the reception are glass-encased meeting rooms and a conference room, their basic white minimalism animated by light fixtures in matte black and hammered copper. The work area is similarly minimalist, with open bench-type desks arranged around the naturally lit circumference of the floor. Enclosed areas were located against the building core.

Descending the staircase's conventional steps or built-in slide brings one into a versatile "café". For private discussions, a circular timber platform with seating for six beckons, partially enclosed with curved panels of fabric mesh. Nearby is a long space enclosed on three sides with cosy, café-like seating. For those who simply need to relax, a series of "daybeds" are located against the window wall, complete with drop-down privacy curtains.

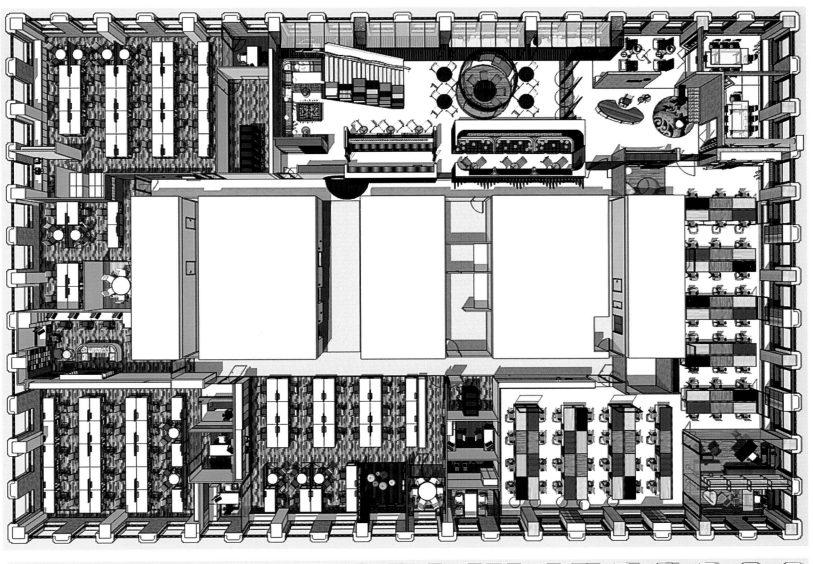
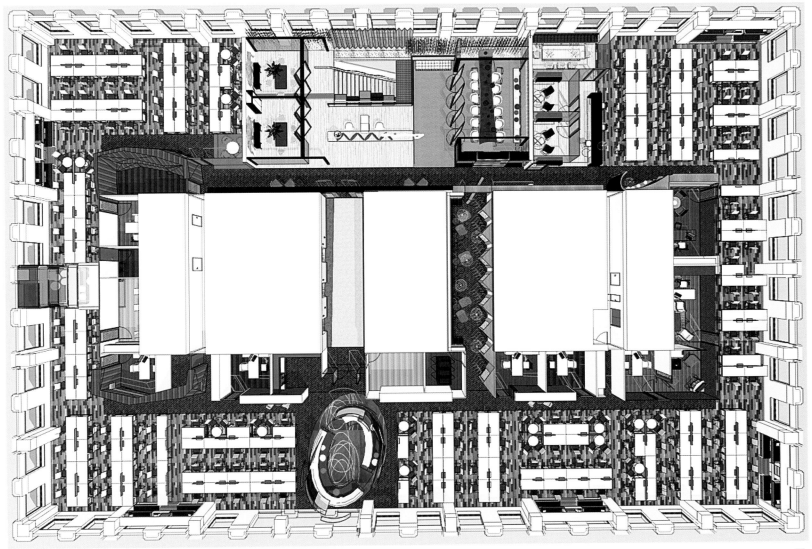

LOCATION Shenzhen, Guangdong Province, China
TENCENT DAZU OFFICE

DESIGNER
edg

DESIGN TEAM
David Ho, Xuanyi Li, Jerry Li, Celline Su, David Gao

AREA
70,000 m²

PHOTOGRAPHER
Sun Zhong Bao

In 2010, Tencent, provider of China's largest Internet service portal, sought to occupy a new office space in Shenzhen that would serve as inspiration for its largely Gen-Y staff, reflect the company's creative culture and display its unique position between physical and virtual reality.

A "regenerative design" approach was used to create Tencent's new interior office space. This approach addresses the issue of environmental sustainability from the perspective of a dynamic organization, allowing for spatial adaptation as organizational changes are required.

Beginning with a detailed discussion about the company's rapid growth, organizational restructuring and future expansion plans, the design team quickly realized the need to design a space with minimal physical boundaries. This meant making necessary partitions and non-load bearing walls less apparent through the use of semi-translucent materials such as tensioned membrane, metal mesh and wooden grills. In addition, breakout and collaborative areas were positioned adjacent to workstations, offering employees flexible working arrangements. To inspire creative freedom, public corridors and private meeting rooms display playful graphic elements, while each lounge area uses a customized palette of bright colors and organic shapes based on each floor's visual theme. To address environmental sustainability, materials such as 100% recycled carpet, glass, and steel grills, as well as rapid growth bamboo flooring were specified, while ambient office lighting was limited to select locations to encourage the use of individual task lighting and increase energy-efficiency. Once construction was underway, a healthy level of indoor air quality was maintained via 24-hour ventilation, whereby all filters, return grills and outlets were sealed to prevent dust penetration into the mechanical, engineering and plumbing system, successfully passing through the design team's eight project management milestones.

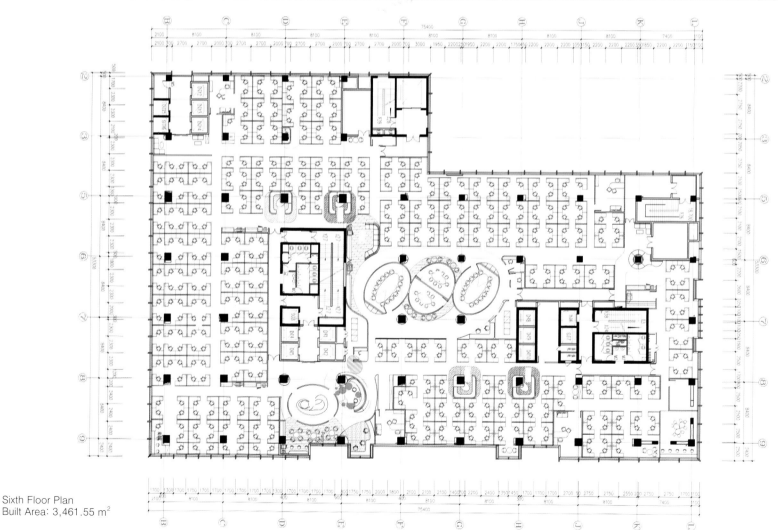

Sixth Floor Plan
Built Area: 3,461.55 m²

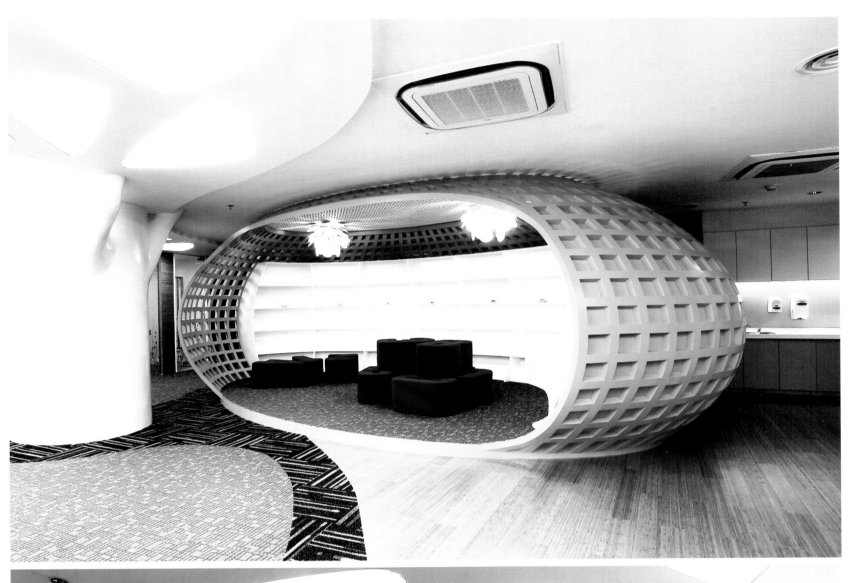

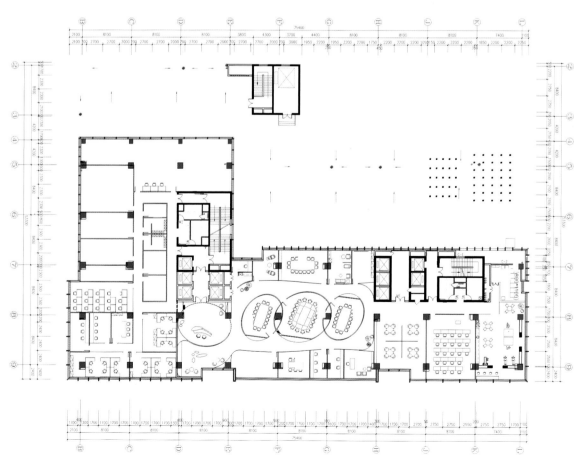

Seventh Floor Plan

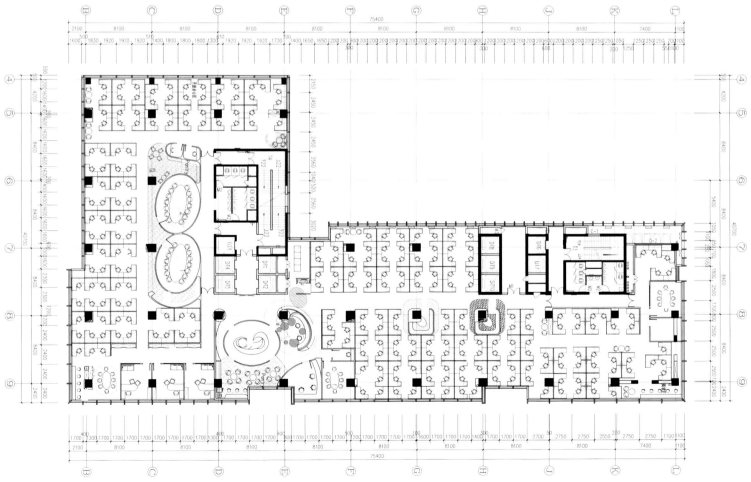

Eighth Floor Plan
Built Area: 2084.76m²

LOCATION Saint Petersburg, Russia

YANDEX SAINT PETERSBURG OFFICE II

DESIGNER
za bor architects

AREA
3,310 m²

PHOTOGRAPHER
Peter Zaytsev

It is indicative that this office is to some extent a return to the roots of cooperation of za bor architects and the largest Russian IT-corporation Yandex.

After a long thought Peter Zaytsev, and Arseniy Borisenko, the project architects, decided to use the double loaded zoning, with meeting "cells", work areas, and unusual objects located along the corridor. As a result of this concept implementation, guests find themselves "inside" the Yandex search service: at the reception they are met by a well-known "Search" button and a yellow arrow (an unofficial Yandex logo and a significant part of the web-site). While passing the corridors they see the familiar user name and e-mail password input boxes, and at each step they meet symbols and icons of Yandex services, although they are not always easy to recognize as tiny pixel icons, had turned into 3D objects.

The project has turned out rather complex, the first thing because no one did such things before, not only in Saint Petersburg but even in Russia. Therefore, many solutions are made on the spot during on-site designer supervision.

As Yandex offices have twenty-four-hours operation schedule, the project was provided with variety of well-developed recreation zones. In addition to working areas and rooms, the office has a gym, cafeteria, showers, and several coffee-points. The number of formal and informal points for negotiation, two lecture halls, and workplaces perfectly equipped with Herman Miller and Walter Knoll systems, make this office a place of attraction, fascinate visitors, and surely makes work very enjoyable pastime.

SKYPE'S NORTH AMERICAN HEADQUARTERS

LOCATION Palo Alto, California, USA

DESIGNER
Design Blitz

AREA
5,017 m²

PHOTOGRAPHER
Hoffman Chrisman, Matthew Millman

Skype's primary goal was to create a world-class office that would differentiate them from their Bay Area competitors in the recruitment of talent. Skype hired Design Blitz architects to design their new North American headquarters on an aggressive budget and schedule.

To really understand how Skype operates culturally, Design Blitz undertook extensive user-group surveying, and researched workspace typology before ever laying pen to paper. A significant portion of Skype's culture is built around Scrum development and a philosophy called "Agile Thinking". To support Scrum, Blitz designed a system of mobile white boards called Skype-its that are distributed throughout the project. The boards can be easily moved and stored depending on a development team's process and requirements. Blitz also created a multitude of different environments to support different thought processes. All of the casual meeting areas are unique and there are three distinct phone booth types: light and bright for active thought; medium colored for meditative thought; and dark cave-like rooms for introspective thought. Like many organizations, Skype required three distinct types of spaces: collaboration, contemplation and concentration spaces.

The building itself provided the greatest source of design inspiration. It was a dark and dingy space with years of tired tenant improvement projects layered on top of one another. The architects made a decision to rip out all the existing ceilings and furring around the steel and never looked back. The resulting space is raw, industrial, and suggestive of a warehouse, which stands in perfect contrast to the highly refined meeting room pods that Blitz inserted into the open space.

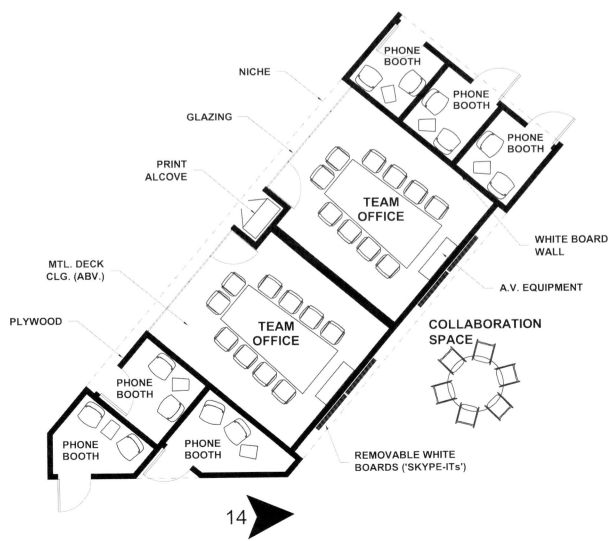

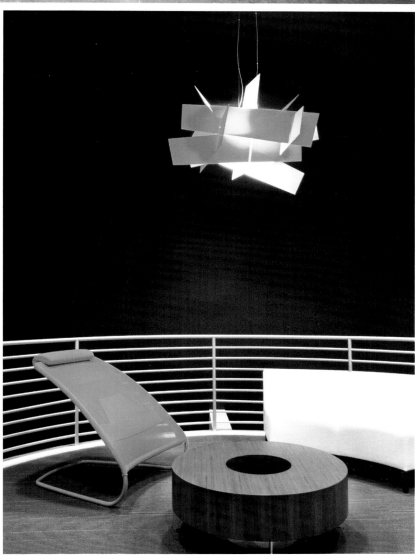

LOCATION Moscow, Russia

CASTROL OFFICE

DESIGNER
za bor architects

AREA
2,025 m²

PHOTOGRAPHER
Peter Zaytsev

Castrol is a global brand and leading manufacturer of engine oils and greases. This British company is a part of BP group of companies.
The office occupies 10th and 11th levels of Paveletskaya plaza business center. It is an open-space with several meeting rooms and cafeterias, separated with dynamic curved volumes. No prominent replanning was projected, but the office extended a little. Renewed open-spaces were filled with dynamic bright objects – curved volumes of meeting rooms, coffee points and cafeterias. It's interesting that two of the conference rooms were separated with a modular wall and can be easily transformed into one with a total area of 54 meters. The office holds 162 workplaces.
The original sectional lamps developed by za bor architects are installed in the main corridors and several conference rooms. Total length of these sections is more than 100 meters. The longest of them is 45m – these unique ad-hoc built-in objects were constructed on-site. Ceiling sections around the lamps were not covered with acoustic material, like in other zones, but were left as is, with communication lines colored in black. This is what emphasized complex curved objects of these giant lamps.
Reception desk was also one of the interesting architectural decisions – due to matte glass and ceramic stone, its outlines seem blurred. Branded prints from race tracks are used as decorative elements. There are also flower parterres and color tiles in the carpet, that increase the idea of liquidity, maintained with the giant lamps.

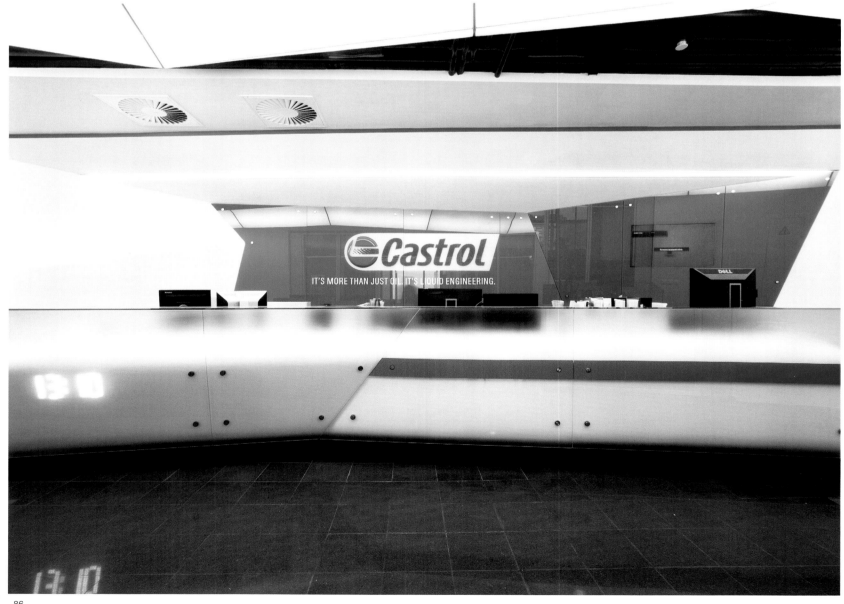

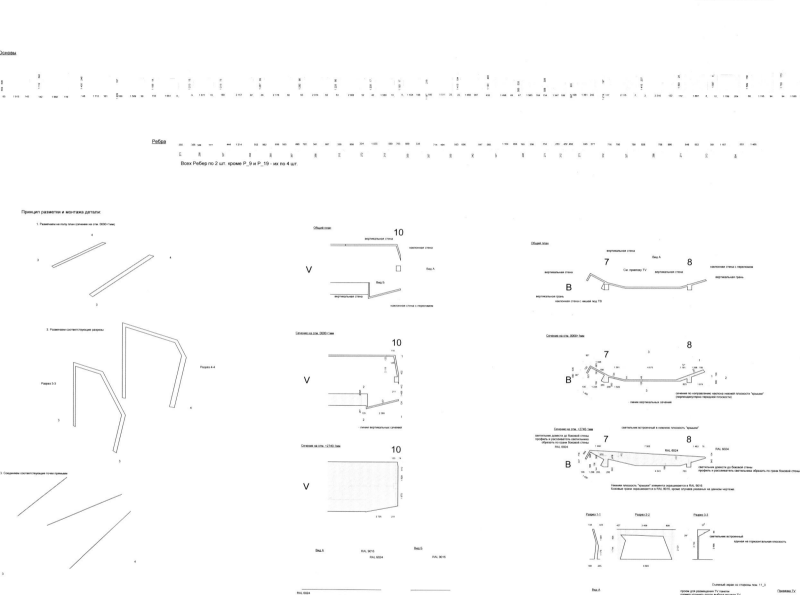

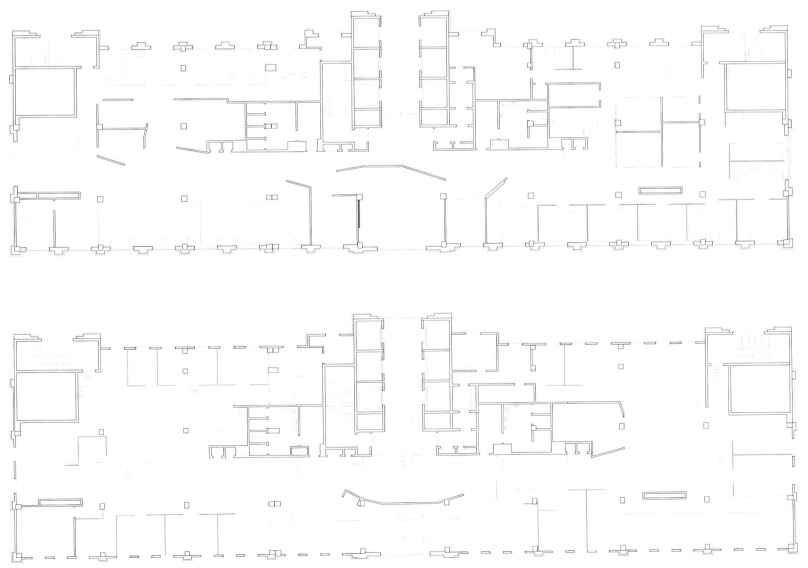

LOCATION London, England

ROPEMAKER

DESIGNER
Clive Wilkinson Architects

AREA
20,207 m²

PHOTOGRAPHER
Riddle Stagg

Macquarie's Ropemaker Place was designed to be a model for a new transparency in banking services, revolving around an open atrium and connecting staircase. Macquarie's aims were to create a unified office environment with a strong emphasis on collaboration, transparency, and client service. Like a vertical "high street", the new Atrium accommodates key communal functions as genuine destinations, which together create a village of opportunities for the workers.

While the Atrium serves to connect the businesses, it also becomes a visual bridge into the workings of the bank for clients and visitors, who are greeted in the ground floor lobby and emerge, via elevators, on the Level 11 Guest Relations area. From here, they can engage in confidential dealings in private rooms, use the conference or event spaces, or meet within the dramatic Atrium volume, enjoying clear views into all the Macquarie workspace. Several sustainable initiatives are achieved with the building, including significant decreases in energy consumption, waste and elevator usage, and net carbon footprint reduction, in line with BREAAM Excellent and LEED Platinum rating. Structural beams cut out of the Atrium were reused to create bridges and cantilevered pods. Engineering cellular space reduces churn costs, and a computer-controlled daylight harvesting system with LED lighting provides significant energy savings. Furthermore, an emphasis on foot traffic using the Atrium staircase has cut elevator usage by 75% while promoting employee health.

In its finished form, the visible energy of people circulating and interacting in the highly transparent Atrium serve as a catalyst for drawing the different business units together and provide the synergy to collaborate with each other in new business ventures. It further underscores the company's brand as an agile, forward thinking and highly collaborative 21st Century Company.

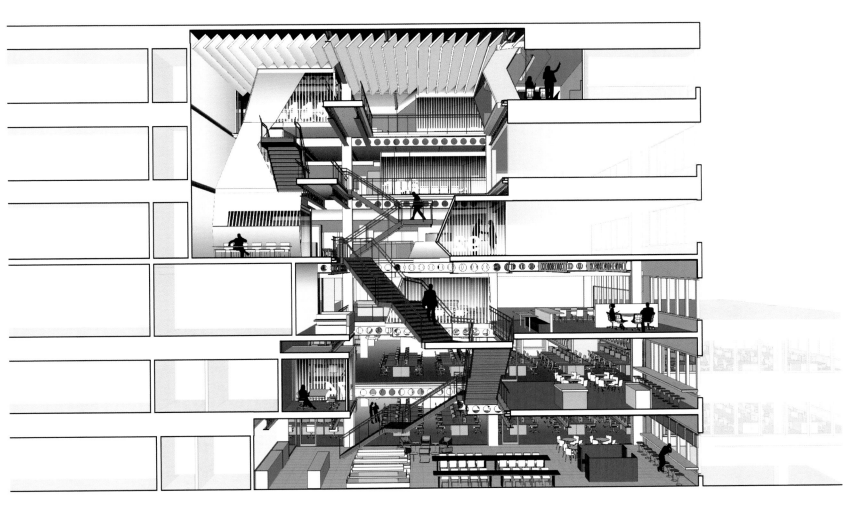

LOCATION Beijing, China

360 HQ OFFICE

DESIGNER
edg

DESIGN TEAM
David Ho, Serena Shu, Tina Ren, Echo Zhang

AREA
18,000 m²

PHOTOGRAPHER
Shi Xiaozhou

With plans to demolish Qihu360's original bland closed-office space, our team was contracted to complete the design and construction of a new "young, free, natural, breakthrough" space which integrates the existing building elements. This project helped to unify and integrate the company's new brand image throughout the space, strengthen the company's corporate culture and enhance its reputation among other Internet-based companies. This project achieved a record of only four months to design and build the 36,000+ m² space. To achieve such a short project schedule, our team made extensive use of factory-made furniture and custom materials which were assembled on site, as well as extensive use of bare roof, open-ceiling plans. As a turnkey project, edg was also able to save time as our team was the designer, contractor and consultant, thus requiring no additional third-party stakeholders. Project teams maintained close communication with the client, to ensure the proper design effect and production cycle are carried through the chain of command.

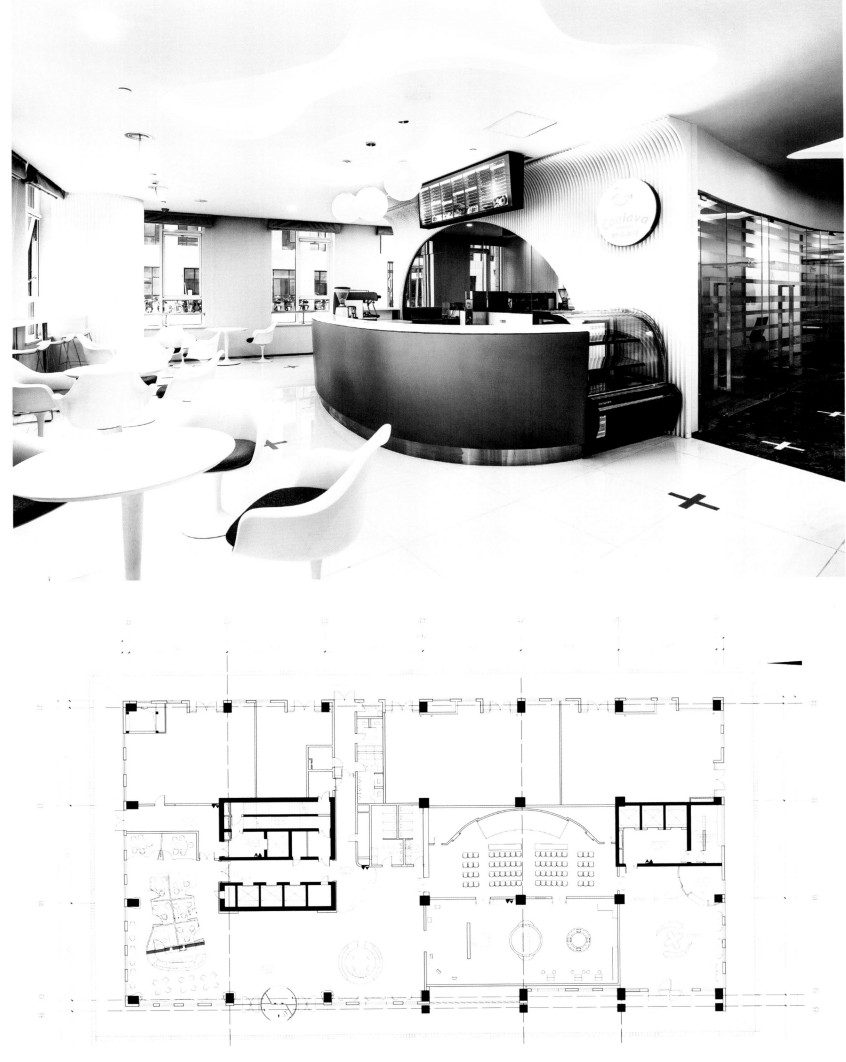

1 Floor Plan

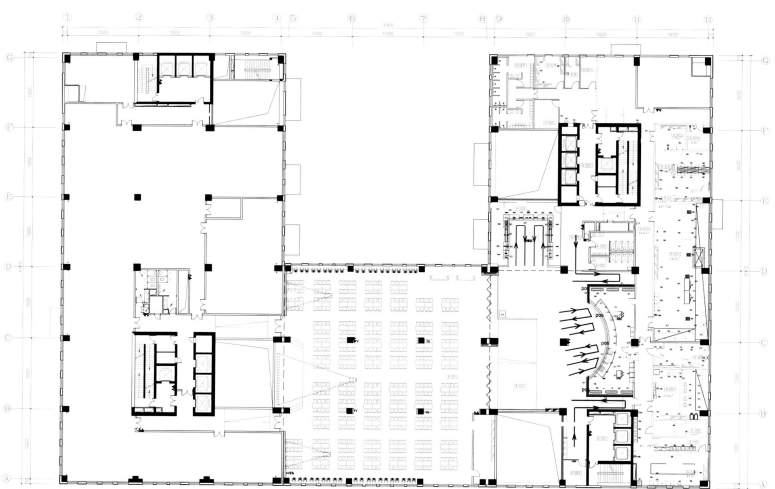

2 Floor Plan

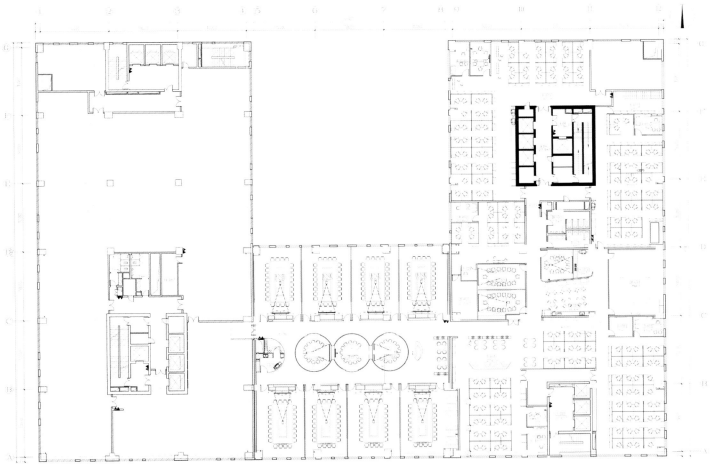

3 Floor Plan

LOCATION London, England

ENGINE INNOVATION LABS

DESIGNER
Jump Studios

AREA
914 m²

Inspired by the amount of research data that planners and strategists of the group's ten individual companies have to deal with on a daily basis and fascinated by different graphical representations of those streams of data, Jump Studios devised an ever changing installation of multi-colored Perspex boxes animated by randomly pulsating LED lights behind.

In response to the brief and to further animate the wall, Jump Studios also designed a series of special units such as two LCD screens which display news feeds, a series of magazine displays and a special light box featuring the Engine logo all of which are integrated in the wall and sit alongside the colored and back lit boxes.

The old reception desk was removed and replaced with a round desk in keeping with a language of white round special furniture items throughout the building.

The second challenge Jump Studios were faced with was the conversation of a 640 m² retail space adjacent to the ground floor reception into an additional office space and a new suite of meeting rooms for workshops and seminars. Innovation Lab One features two flexible clusters of low seating which can be rearranged to suit different numbers and sizes of groups. The last of the three rooms, Innovation Lab Two, is yet another space.

A connective element between all three rooms is a series of identical display screens facing the street facade. These elements play with notions of transparency and opaqueness, concealing and revealing.

The open buffer space with adjacent tea point situated between the three labs and the open plan office offers additional informal meeting opportunities and can be used for lunches and receptions as well.

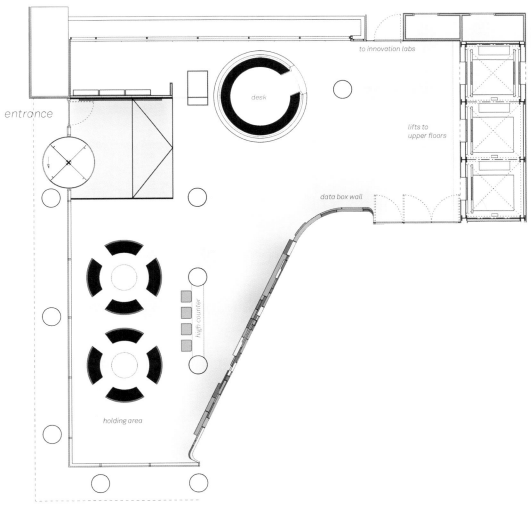

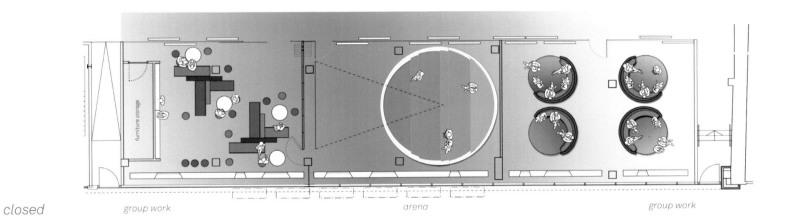

closed   group work   arena   group work

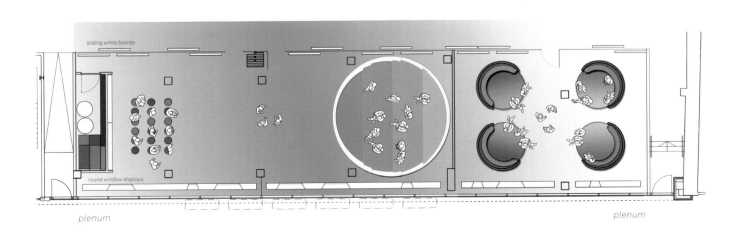

open   plenum   plenum

LOCATION London, England

GOOGLE CSG SUPER HQ

DESIGNER
PENSON

AREA
14,865 m²

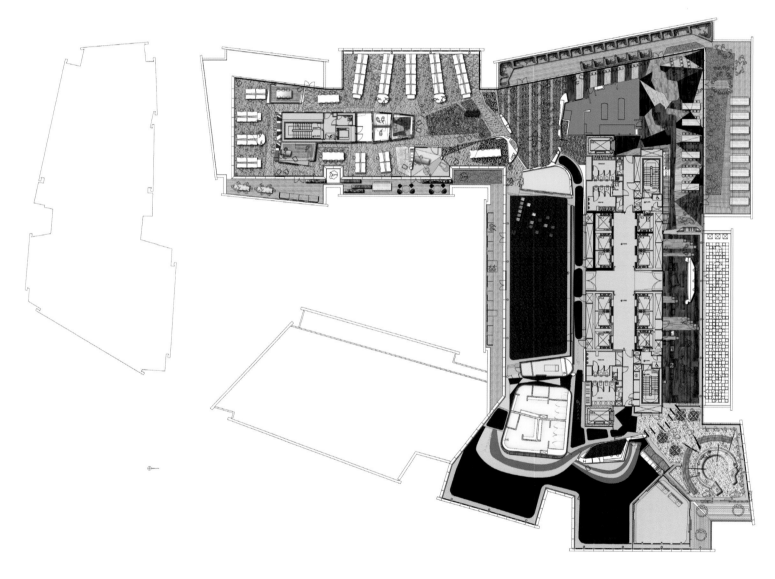

One of Europe's leading interior design and architecture studios PENSON has delivered Google's new super HQ at Central Saint Giles, Covent Garden, London.

The Secret Gardens are accessed from the main café space through little concealed garden gates. They also connect with Google Park, which is a huge garden area also on a sun-trapped balcony. The café again looks non-commercial; it sucks in sky blue from above via water based paint to ceilings, draws in light and fun atmospheres from the park outside.

The Google Green is an interior space which looks out across the Secret Gardens and joins the café, Town hall and small desking area. This space is flexible, as it can be used for "All Hands" meetings, one-to-one slouching in cool Moroso couches, or can be used to entertain large groups of Town Hall guests whilst on their lunch or coffee breaks.

The Town hall has a capacity to showcase two hundred guests in a fully velvet curtained hall, with open exposed ceilings, a massive Video Wall, amazing acoustics and hi-fi capabilities. This space has been placed in the heart of the floor connecting to the Main Reception with a gallery of artwork and trinkets. Glittery walls splashing huge Union Jacks also help to celebrate the UK whilst adding an economical finish that looks amazing whilst being water based.

The Gymnasium has some of the best views across to the south, and provides very quirky shower, change and massage facilities, with a cool Bikedry which provides hanging space for bike gear to air and dry for one's return commute. A dance studio not only assists in flexibility but also in Googlers health and fitness at work. The shower block is painted in an incredible illuminous orange paint that whacks you with a slap of energy.

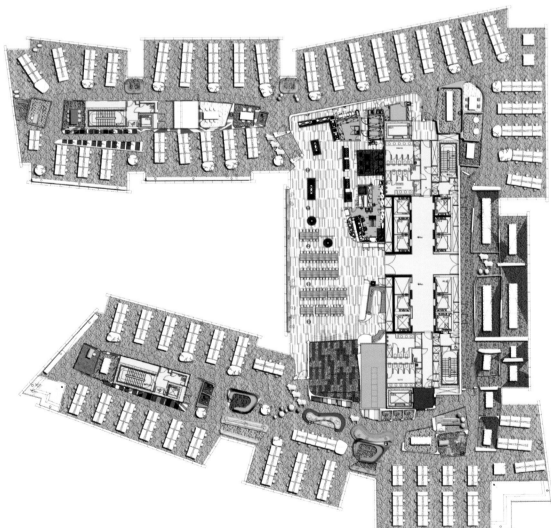

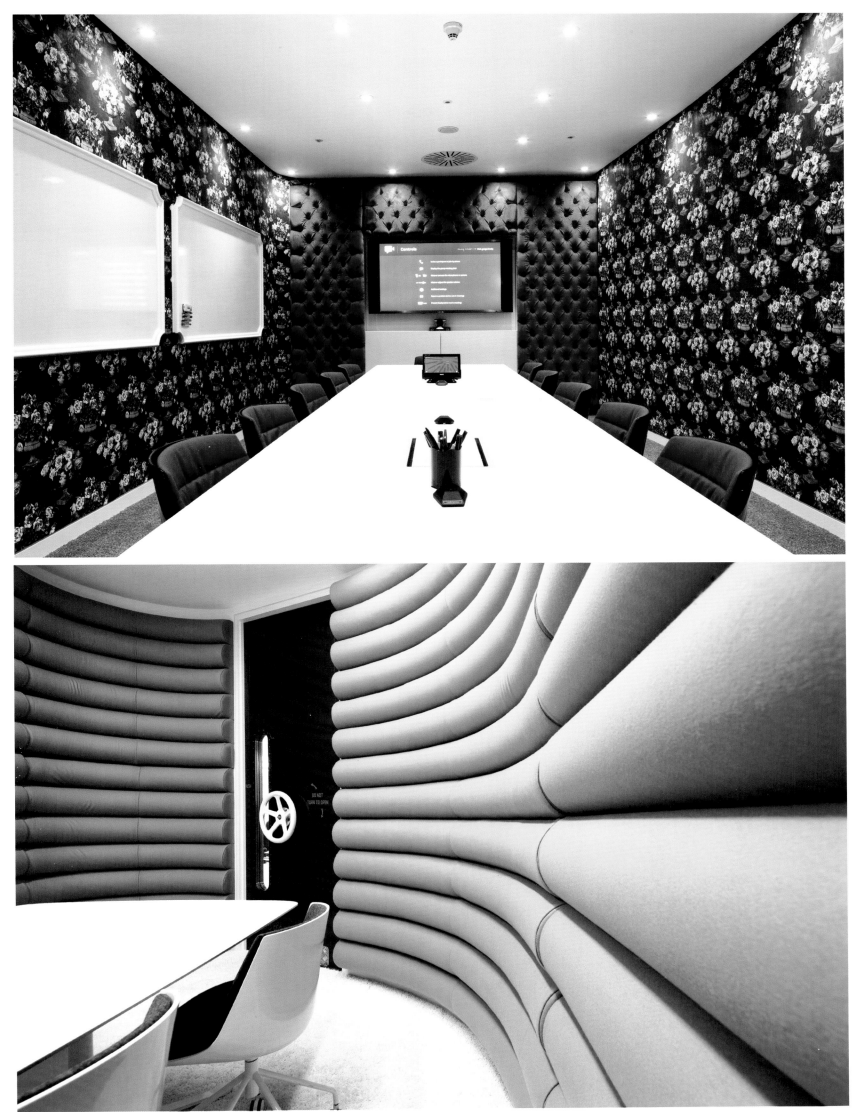

LOCATION Switzerland

UNILEVER OFFICE

DESIGNER
Camenzind Evolution

Camenzind Evolution's concept created a new workspace layout to support the Unilever values and visions in the context of Agile Working. It was essential to transform existing traditional workspaces into flexible space designed to drive innovation and ideas sharing. This was achieved by distributing differentiated functions along the periphery of the building, and additionally creating an open, centrally located staircase linking all three floors for visual and social connectivity. The new open-plan office creates a greater sense of a single team working together, and allows for people to choose where they sit, depending on the team they are working with on any given day.

In the Focus Zones, the goal was to create well-designed and innovative spaces that motivate and inspire employees to excel in their daily work. Within this zone, designated Quiet Areas offer a diverse range of work areas with different moods and atmospheres, such as the cozy and soothing "Library" or energizing "Garden". This creates a choice for the employees depending on their preference and increases their feeling of wellbeing.

The fresh and vibrant "Deli-Bar" and the warm and inviting "Coffee Hub" in the Vitality Zone introduce new social spaces on the building's two main floors. Their central location makes them an informal alternative to meeting rooms, for collaboration, networking and creative brainstorming. Dedicating special areas for employees to revitalize and relax also brings to life for employees Unilever's commitment to helping people "feel good, look good and get more out of life".

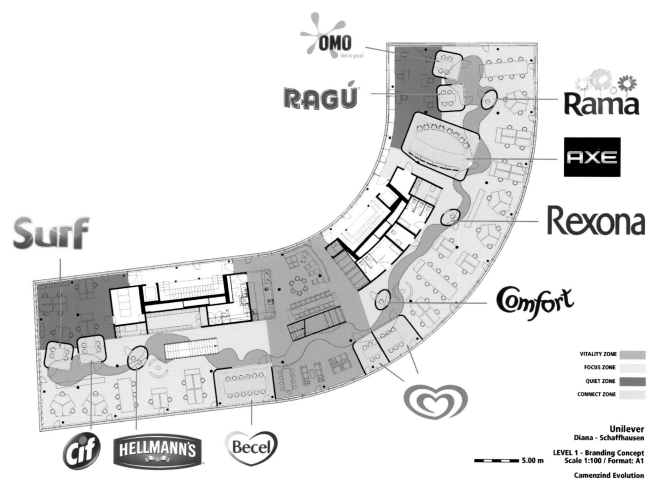

VITALITY ZONE
FOCUS ZONE
QUIET ZONE
CONNECT ZONE

Unilever
Diana - Schaffhausen

LEVEL 1 - Branding Concept
Scale 1:100 / Format: A1

Camenzind Evolution

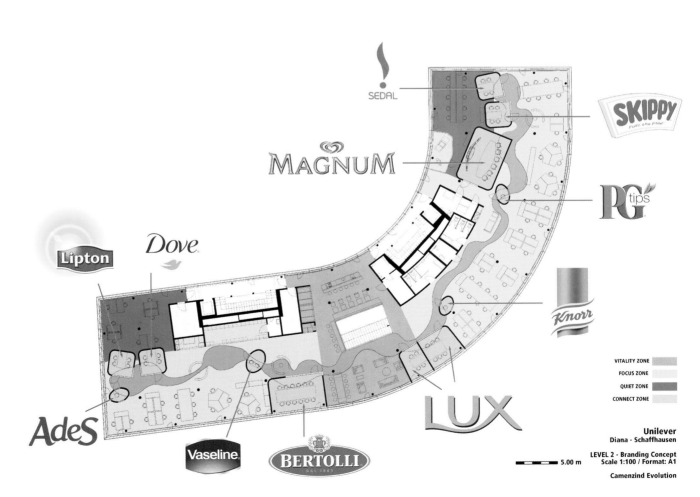

Unilever
Diana - Schaffhausen

LEVEL 2 - Branding Concept
Scale 1:100 / Format: A1

Camenzind Evolution

---LOCATION--- Bogota, Colombia

JWT

DESIGNER
AEI

AREA
2,300 m²

PHOTOGRAPHER
Andrés Valbuena

The project developed by Aei is located at Cll 96 No. 11B-56 in the fifth floor to the eight and terrace. The terrace of approximately 300 m^2 has the main cafeteria and the general dining room can be used to develop social activities, meetings and exhibitions of large-scale. However, within the same spaces the project provides informal spaces for smaller meetings. Inside, offices depart fundamentally from a computer principle clear, given by two orthogonal and intercepted major circulations complemented, in the majority of cases, with open meeting areas, whose location was proposed from the conception of the project, understanding the needs of the client.

The space required for JWT is from these circulations. On each floor, areas such as the points of copied, coffee points, and boardrooms are warranted to allow less movement of the staff, and thus contribute to efficiency in the daily activities of the company.

The image of the offices can be summed up within the precepts of sophistication, personality and uniqueness. Also was proposed the creation of a fixed point inside, a staircase that provides a focal and blunt object to the project. That element was thought in red, being the sole purpose of color intense in the project, which can be seen from almost anywhere. The tones of the finishing of floors, furniture and fixtures are mostly neutral colors where the importance leads it to the texture of the material. Floor mat traps dirt, wood furniture in pine, sofas in leather texture and linings of walls in brick, lamps with glass elements, brown, gray, black and white colors in the furnishing, are the results of previous architectural reflection.

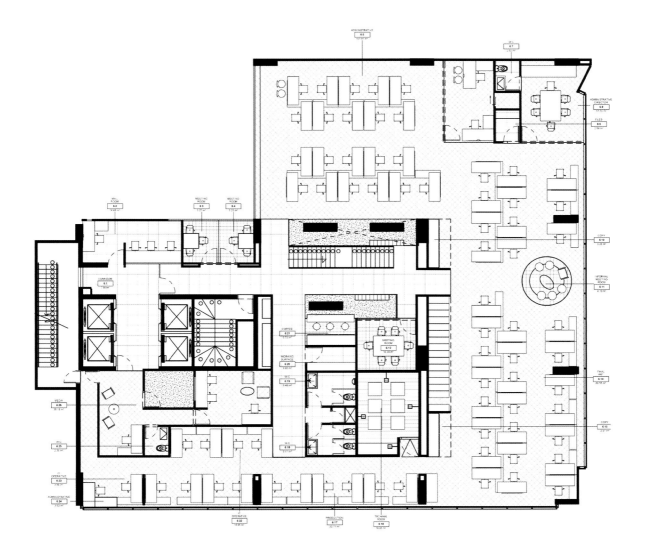
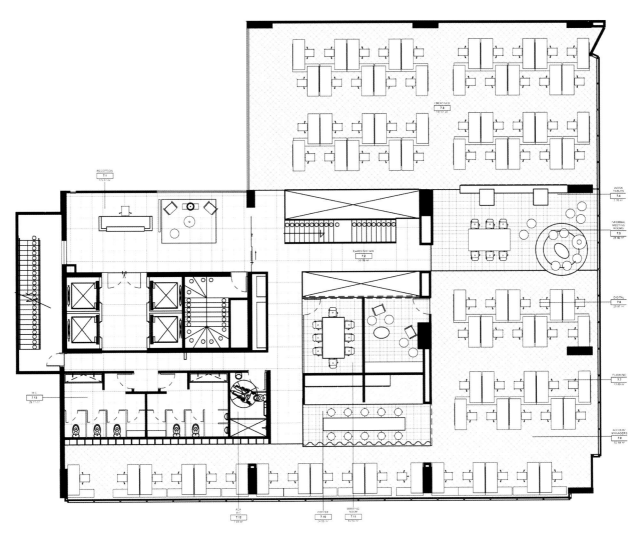

LOCATION   Amsterdam, the Netherlands

RED BULL

DESIGNER
Sid Lee Architecture

PHOTOGRAPHER
Ewout Huibers

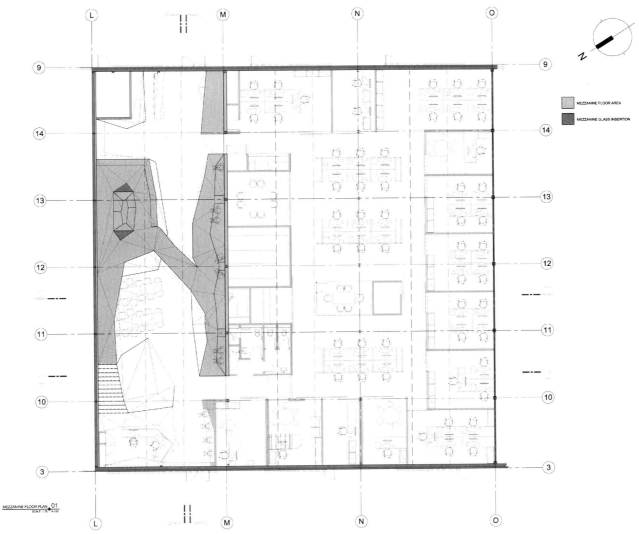

MEZZANINE FLOOR PLAN 01

Chosen over two other firms, Sid Lee Architecture and Sid Lee's Amsterdam atelier were mandated to create the new Red Bull Amsterdam headquarters. The company agreed to settle in the North side of Amsterdam's port area, in a site evocative of both an artistic street culture and the intensity of extreme sports. The project landed in an old heritage shipbuilding factory, facing a timeless crane and an old disused Russian submarine.

"To design the inner space, we aimed at retrieving Red Bull's philosophy, dividing spaces according to their use and spirit, to suggest the idea of the two opposed and complementary hemispheres of the human mind, reason versus intuition, arts versus the industry, dark versus light, the rise of the angel versus the mention of the beast", says Jean Pelland, lead design Architect and Senior Partner at Sid Lee Architecture. Inside the shipbuilding factory, with its three adjacent bays, the architects focused on expressing the dichotomy of space, shifting from public spaces to private ones, from black to white and from white to black.

Our goal in this endeavour was to combine the almost brutal simplicity of an industrial built with Red Bull's mystical invitation to perform. The interior architecture with its multiple layers of meaning conveys this dual personality, reminding the user of mountain cliffs one moment and skate board ramps the next. These triangle-shaped piles, as if ripped off the body of a ship, build up semi-open spaces that can be viewed from below, as niches, or from above, as bridges and mezzanines spanning across space. In the architecture we offer, nothing is clearly set; all is a matter of perception.

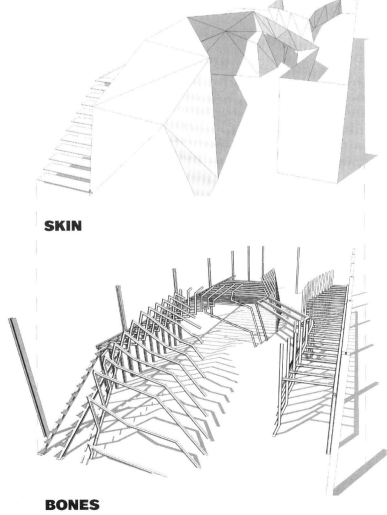

SKIN

BONES

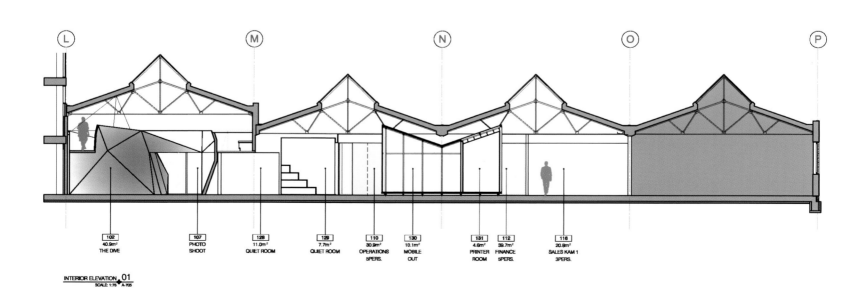

INTERIOR ELEVATION 01
SCALE 1:75   A-705

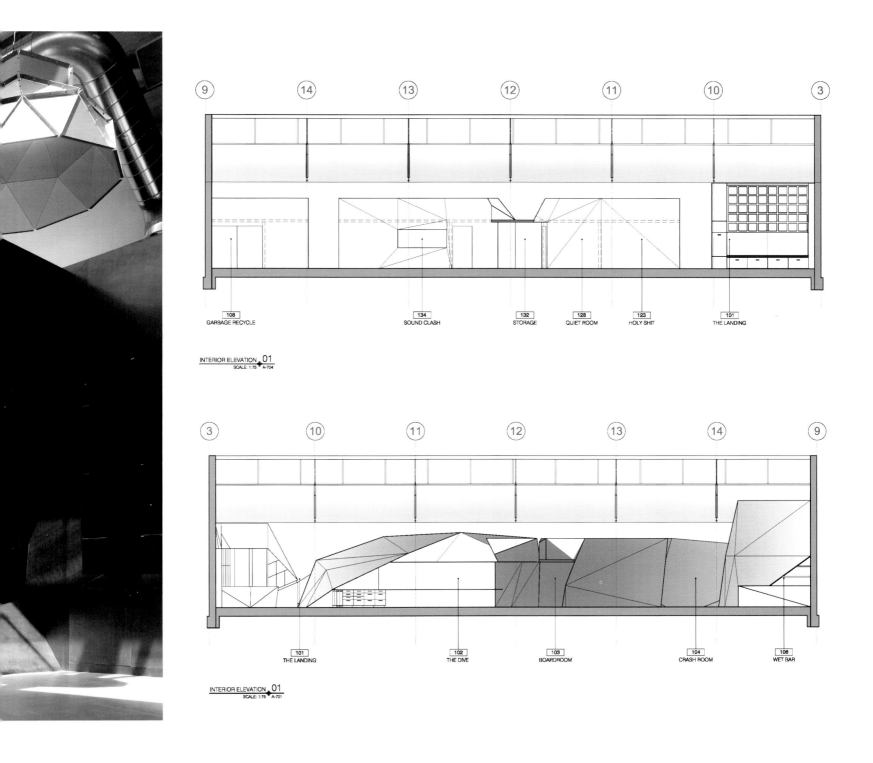
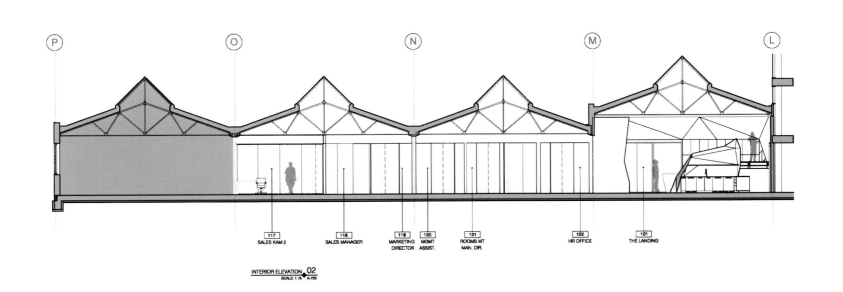

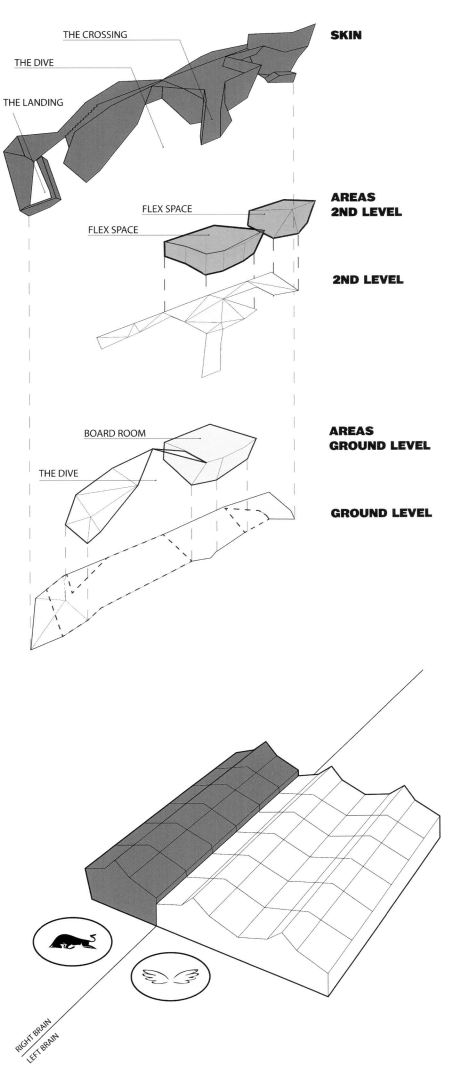

151

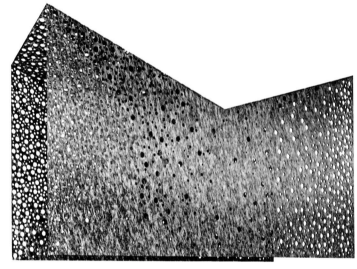

METAL SHELL

GLASS ENCLOSURE

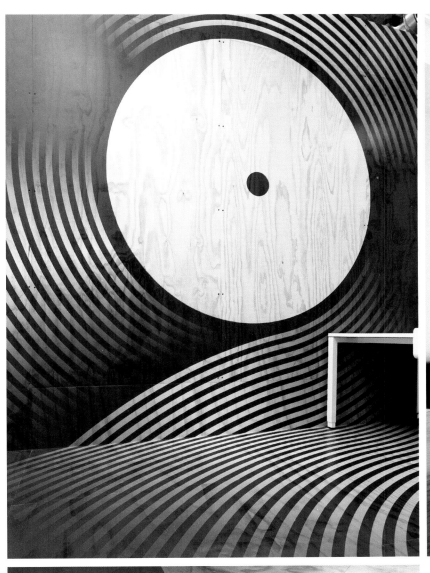
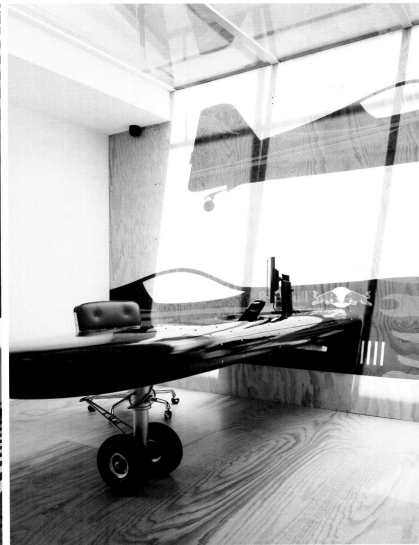

LOCATION Arnhem, the Netherlands

SIDN

DESIGNER
D+Z Architecten+Projectmanagers

AREA
2,300 m²

As the administrator of the Netherlands, SIDN is responsible for the functional stability and development of the Netherlands' country-code domain. SIDN registers the Netherlands domain names and ensures that registered domains – of which there are now more than five million – remain reachable on the Internet. Every day, SIDN handles more than a billion search queries from Internet users, enabling them to reach websites and e-mail addresses.

SIDN asked D+Z to design and realize the office interior when they relocated to a new office in Arnhem. Aize Oenema, the interior architect, had an extensive "DNA-profiling" session with SIDN before commencing with the design of the interior. From this "DNA-profiling" Aize learned that SIDN would benefit from introducing the New Ways of Working in their new office. He implemented this in the design and assisted SIDN with the introduction of this flexible working concept.

The result is a modern, colorful interior as transparent as the Internet itself.

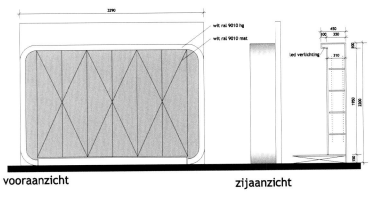

vooraanzicht   zijaanzicht

bovenaanzicht

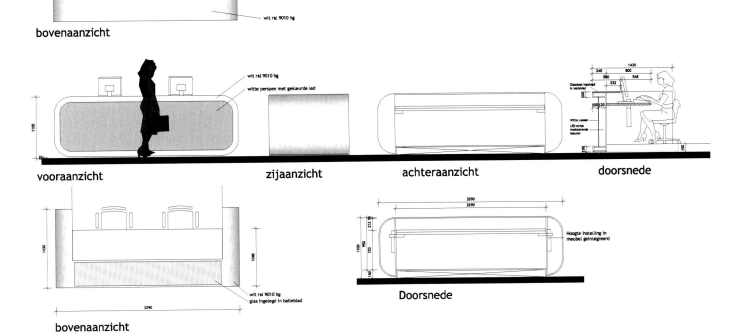

vooraanzicht   zijaanzicht   achteraanzicht   doorsnede

bovenaanzicht   Doorsnede

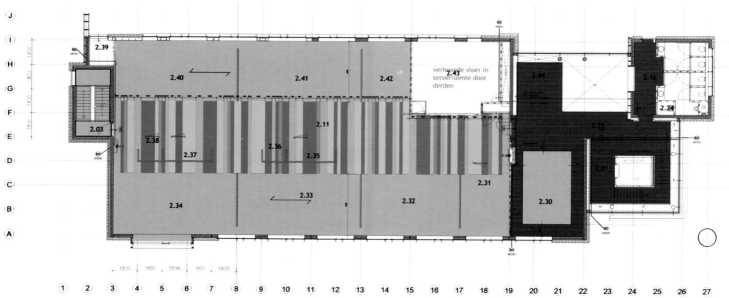

Legenda vloerafwerking

■ Hout ntb
■ tapijt tegel Desso Flux A786 9095
■ Desso Palatino A072 4007
■ Desso Palatino A072 3118
■ Desso Palatino A072 8921
■ Desso Palatino A072 7218
■ Desso Palatino A072 8406
■ Desso Palatino A072 8431
Forbo marmoleum Walton kleur 171 (cement)
Forbo colovinyl kleur 989 (onyx)
Tegels Mosa, Terra Maestricht kleur 227V, afm. 900x900mm
schoonloopzone Coral Classic kleur antraciet
▨ vloerkleed op harde ondervloer Object karpet, type Tosh, kleur 1403

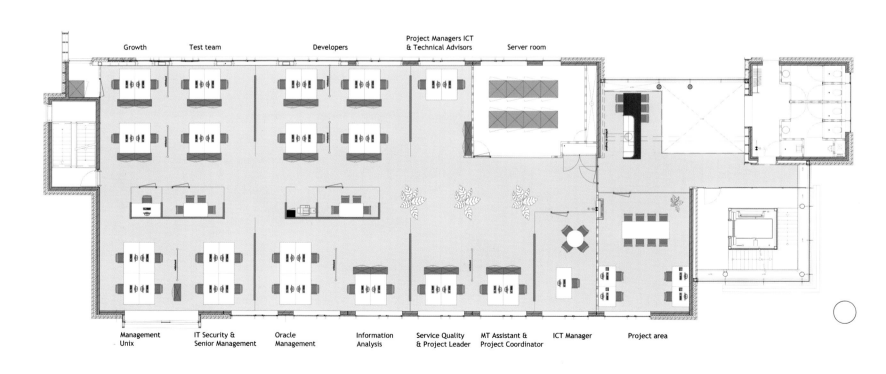

| Growth | Test team | Developers | Project Managers ICT & Technical Advisors | Server room |

Management Unix | IT Security & Senior Management | Oracle Management | Information Analysis | Service Quality & Project Leader | MT Assistant & Project Coordinator | ICT Manager | Project area

LOCATION London, England

ENGINE OFFICES, LONDON

DESIGNER
Jump Studios

Jump Studios has designed an animated but suitably slick new office for London-based communications group Engine.

With 12 different companies operating under the Engine umbrella, the challenge was to create an environment that would appeal to a broad range of tastes while respecting and upholding the individual brand identities.

The team worked with the concept of "precision engineering", partly inspired by Engine company chairman Peter Scott, known to be a "razor sharp" businessman with a passion for detail. "We really wanted the office to look like it was machined rather than constructed," Jordan explains. This idea is most clearly manifest in a series of perfectly formed elements that run through the building, essentially forming a backbone that links the ground floor to the fifth. One of the most dramatic is the floating auditorium at entrance level, designed for presentations. "As a sculptural element I think this works pretty well," says Shaun Fernandes, who led the design team. "And of course it's a talking point for clients."

What among many other "talking points" in the building are the seating pods on the fifth floor with their Corian shells and Barrisol light ceilings. Here employees are encouraged to interact, serviced by a café offering spectacular views across the city's rooftops and a series of conference and meeting rooms ranging in design, size and style. Nonn explains: "We wanted to offer a more refreshing alternative to the conventional table and chair." Some of the more imaginative solutions include "mini auditorium" seating systems and one room clad entirely in cork allowing for quick, non-permanent customization.

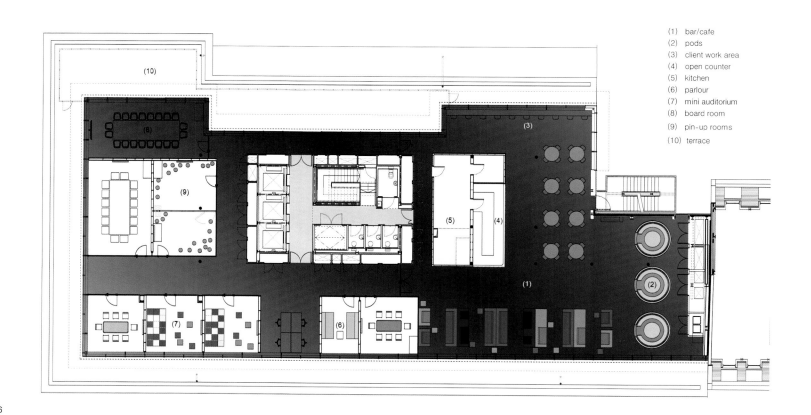

(1) bar/cafe
(2) pods
(3) client work area
(4) open counter
(5) kitchen
(6) parlour
(7) mini auditorium
(8) board room
(9) pin-up rooms
(10) terrace

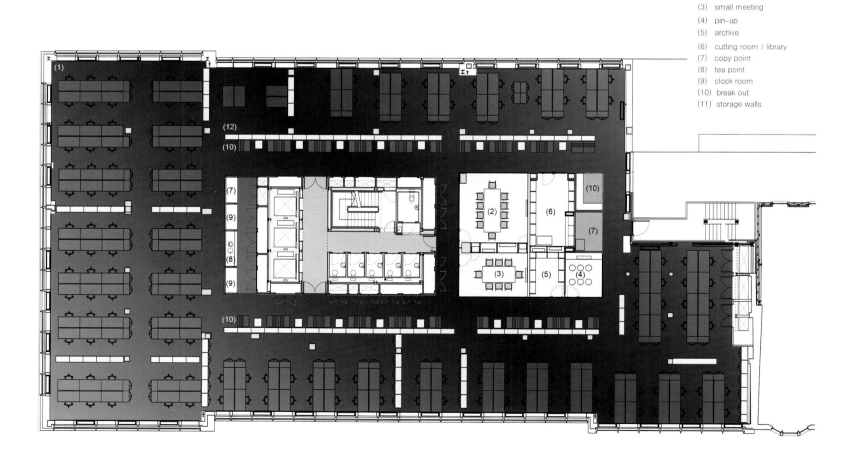

(1) open plan office
(2) large meeting
(3) small meeting
(4) pin-up
(5) archive
(6) cutting room / library
(7) copy point
(8) tea point
(9) clock room
(10) break out
(11) storage walls

LOCATION Warsaw, Poland

B+L HQ

DESIGNER
137kilo, Beza Projekt s.c.

The project is the Poland headquarters of a large pharmaceutical corporation. The client wanted the office space to highlight the company's scientific heritage while providing a high quality working environment for its employees.

Research labs and contact lenses – the client's main product – served as our inspiration. White exposed ceilings combined with glass surfaces combine to create a light and transparent space. This seemingly sterile aesthetic is balanced by green walls along the space's main visual axes and functional solutions which encourage teamwork and relaxation.

An open-space arrangement was used for many of the workspaces. The open space is punctuated with meeting room domes which also delineate the office's various departments. The domes are meant for less formal work meetings. Their high-tech inflatable surfaces invoke the client's main product – contact lenses. The domes' internal steel structure provides rigidity and adherence to fire codes. The domes are connected to ventilation and air conditioning systems – creating a comfortable work environment.

The client's desire to provide a comfortable work environment is manifested by the "fun room" which features a library, a football table and an adult-size playpen. Work spaces are some of the project's key elements. Each four-person work area invokes the "+" symbol (an element of the company's logo) and has a small tree in its center. Employees at each area are responsible for watering their tree, which underlines the company's commitment to team work. The plants in the green walls were selected for their ability to purify indoor air and absorb dust. The design was tested on scale models – the final version was presented to the entire team before the start of construction.

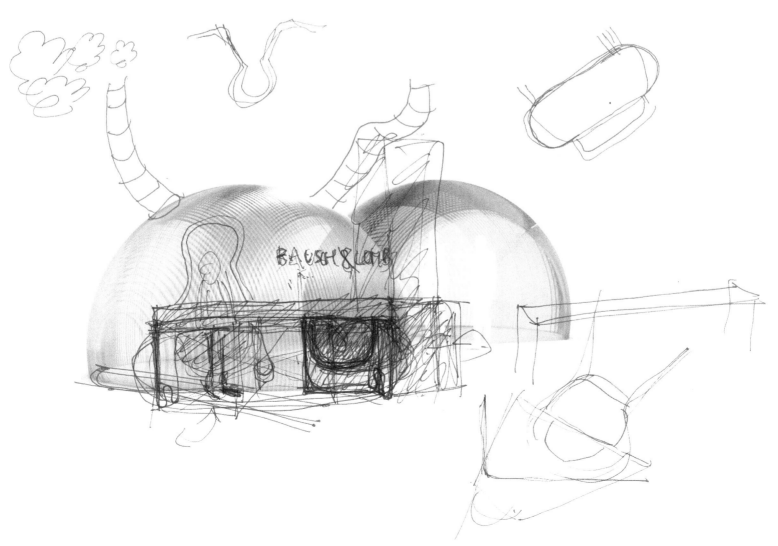

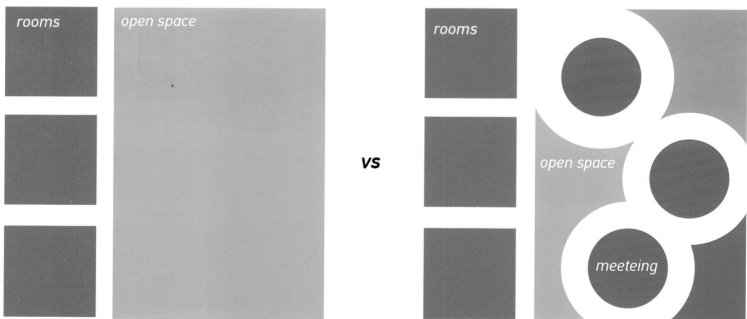

181

LOCATION London, England

GOOGLE L3

DESIGNER
PENSON

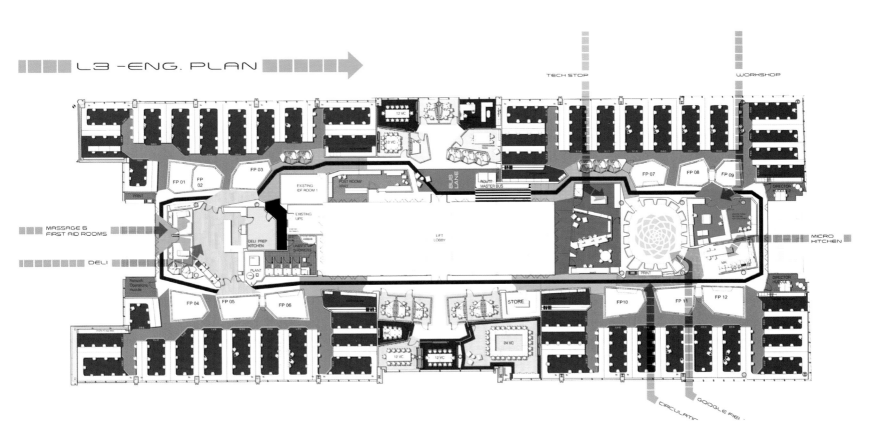

L3 – ENG. PLAN

Leading global architecture & interior design agency PENSON, has released images of their latest revolutionary Headquarters for Google, located in Victoria, London.

Just take a peak into the Deli, which serves up a great combination of natural materials to compliment the healthy food. The Flight Pods seen in some previous PENSON schemes have been developed a little further, they are lighter, yet more acoustic and provide more fun internally with more white-board externally. They even accommodate a LED monitor for reporting important information to Engineers on a constant basis. These flat pack rooms as per usual provide a multi-function space that can be used for meetings, brainstorming, war-rooming or Friday pizza consuming.

Other spaces such as the workshop continue to recycle & thrive on Google's Red List of healthy ingredients in terms of building materials & finishes. Each space is different yet is linked by some very cool circulation routes that always provide inherent opportunities for collaborating, grabbing a print you've sent or grabbing a vista into a cool nook & cranny to brighten up one's day. PENSON have used recycled seat belts & scaffold poles to create "not-seen-before" details to economically design-up these routes. There is also a naked "exposé" theme running through the spaces, where the workings of the HQ can be seen through glimpses of linings is missing. This occurs to some table tops also, with little trinkets beneath to add a bit of zest to one's video conference.

What with the cool Micro-Kitchens, wide variant of meeting room sizes, shrewd space-planning & overall interior design, is a great HQ and this underlines PENSON's standing as one of the world's best workplace architects.

L3 - ENG. VISUAL PLAN

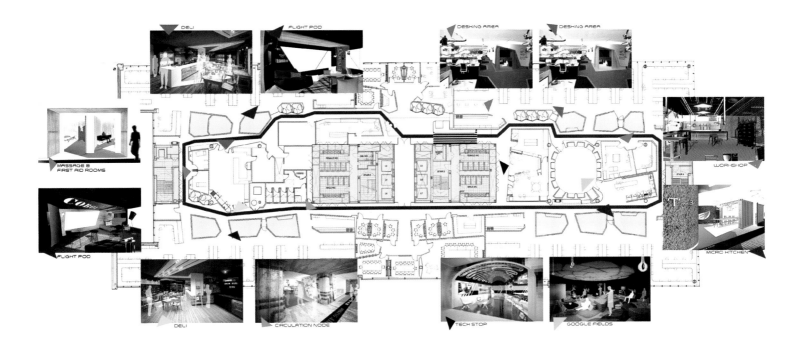

LOCATION San Francisco, California, USA

HEAVYBIT

DESIGNER
IwamotoScott Architecture

Heavybit, a new, curated community for cloud developers, is designed as a series of architectural interventions inserted into an existing three-storey warehouse shell. The interventions define space, accommodate program and work materially with the client's concept of heavy physicality coupled with the ephemerality of the cloud.

Given the project brief and a limited budget, the program is addressed through a series of designed interventions inserted into the existing shell. The largest of these is a multi-functioning platform at ground level with a new stair leading to the second floor. The platform, constructed as a "solid" laminated plywood object houses the reception desk located opposite the main entry, bar-height work counter that doubles as seating on the raised platform side, speaker stage facing the dining area, pass-through ramp, and U-shaped lounge seating. From the rear it appears as a series of steel plates, from the side it almost disappears. On the second floor, the fins extend upward to become a guardrail "box" for the stair, with a bar height work-shelf on one side. The ground floor conference room and bike room are defined by a gray, reclaimed wood and glass wall, and the kitchen is made of dark gray stained plywood. The second and third floor workspaces are planned as an open environment with several conference rooms on each floor. The conference rooms are made to appear as a set of sliding walls constructed of Polygal over steel frame and painted exposed metal studs.

Lastly, three smaller interventions were commissioned on a design-build basis. These include the rope room, and also "hex cell light" over the kitchen island, and "hex cell ceiling", a fabric light diffuser in the first floor conference room.

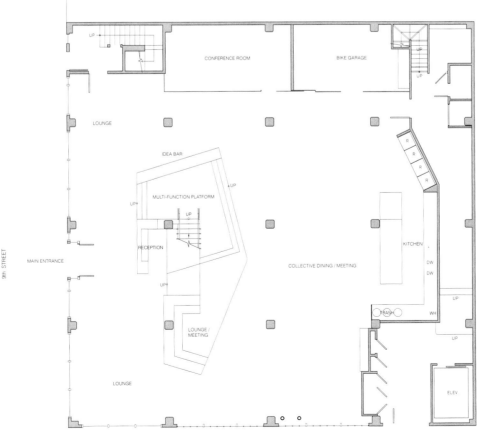

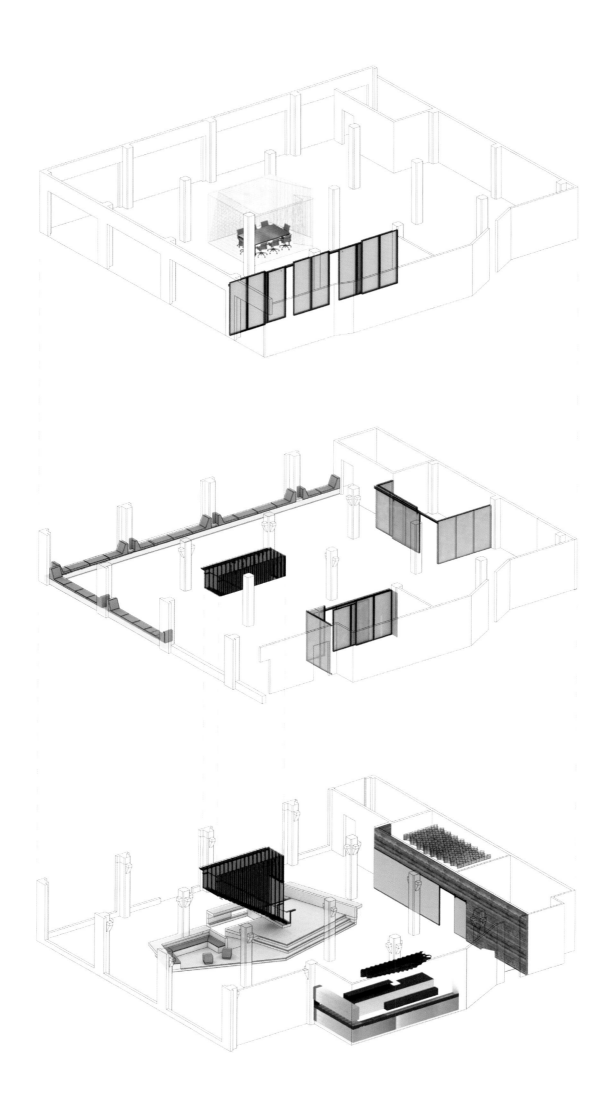

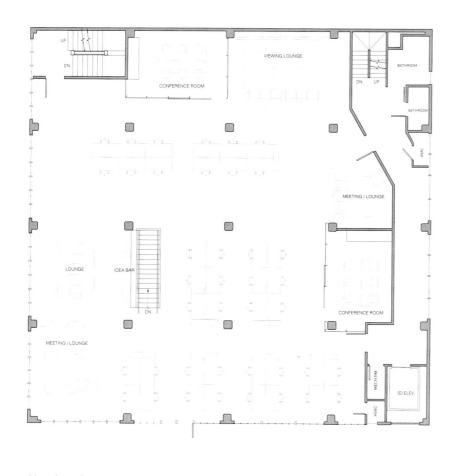
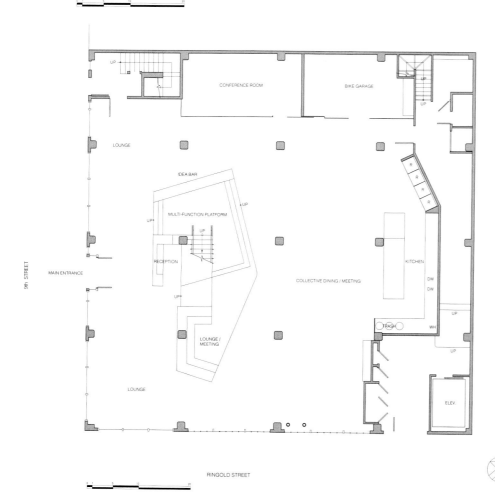

LOCATION: Vienna, Austria

HQ SAMSUNG VIENNA

DESIGNER: INNOCAD
AREA: 4,000 m²
PHOTOGRAPHER: Paul Ott

The task in this project was to create an inspiring, communicative and functionally optimised work environment within the existing office premises of Samsung Austria.

The prerequisite was that today's knowledge workers require conditions that permit dialogue, meeting and networking but also peace and quiet and concentration. The aim of the concept is to encourage interaction and thus informal communication among staff and to provide a pioneering communication infrastructure; also, to develop a unique corporate architecture for Samsung Austria.

The office unit spanning five storeys in Vienna's Galaxy Tower was completely reorganized. These classical work areas were combined in a compact unit on four storeys and the middle storey was turned into a central staff communication and collaboration platform. At the same time, these areas are also used as an imposing welcome area and for showcasing. These meeting and relaxation settings appointed with modern furnishings (e.g. cafeteria, lounges, shared kitchen, playground, retreat, focus, phone booth and various rooms for meetings) are in constant use by staff and customers alike and encourage spontaneous exchange, social contact, and thus identification with the company. Key topics from the staff survey GWP (Great Place to Work) were taken into account and incorporated into the new office design.

The new, inspiring work environment encourages creativity and allows every member of staff to use the ideal work and communication environment for any particular task and thus to unfold his or her maximum potential. The new office thus leads to a marked increase in productivity, work-life balance, staff satisfaction, and thus the overall corporate culture of Samsung Austria.

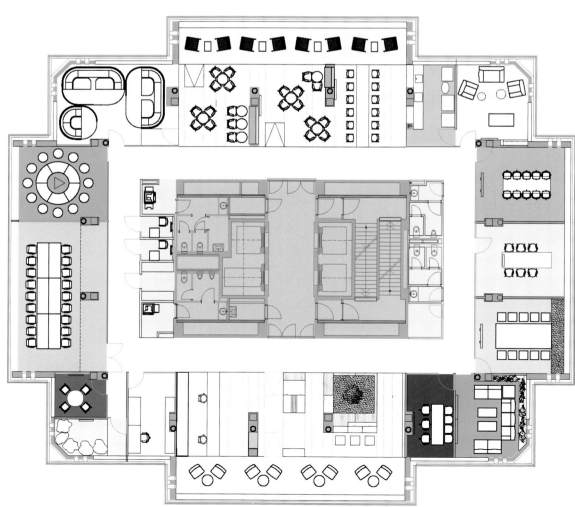

LOCATION Wallisellen, Switzerland
MICROSOFT BRIEFING CENTER

DESIGNER
COAST Office

PHOTOGRAPHER
David Franck

With its design of the Microsoft Briefing Center in Wallisellen, Switzerland, the Stuttgart-based agency COAST Office successfully created a new interactive product presentation and event space that exhibits a high degree of usage flexibility. The challenge was not to display the products in a conventional shop setting. Importance was attached instead to integrating Microsoft technology into the space and to using it for interaction and product presentations: through the media integrated in the furniture and walls, for instance, motion sensors and touch technology is used to control product presentations and content and to illustrate the technology to visitors. Intuitively, customers and employees thus use the various interlinked communication solutions.

Their design aimed to achieve a firm focus on media content, products and direct communication with both customers and visitors. The idea was to create a dynamic, technological and immaterial space which is brought to life alone by the color and content of the media and by Microsoft's customers, staff and visitors. The color scheme of rooms of the Microsoft Briefing Center is therefore strictly black and white. The space is structured by wave-like furniture units which divide the room into a lounge at the front and a conference and presentation room towards the rear. The drapes around the room form a deliberate contrast to the otherwise hard technical surfaces. Moreover, having two layers of drapes in different shades of gray and differing degrees of opaqueness makes it possible to close off or open up the rooms to accommodate the needs of an event. In this way, the color, brightness and mood of the room can also be changed and they have a positive impact on the acoustics of the room.

CONCEPT_FLOORPLAN

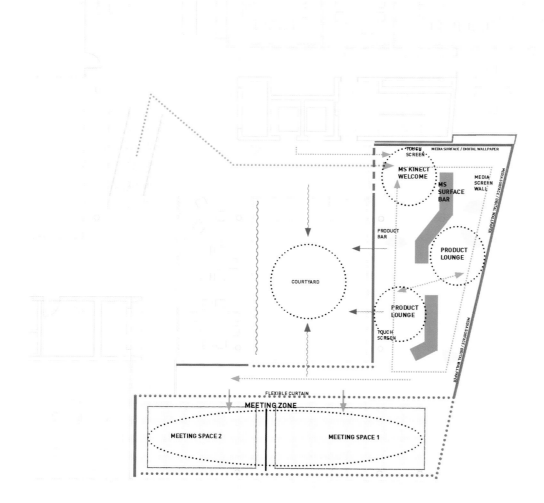

FLOORPLAN

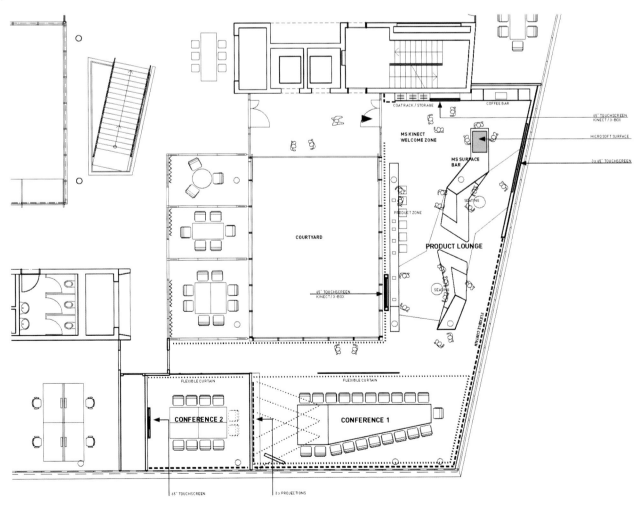

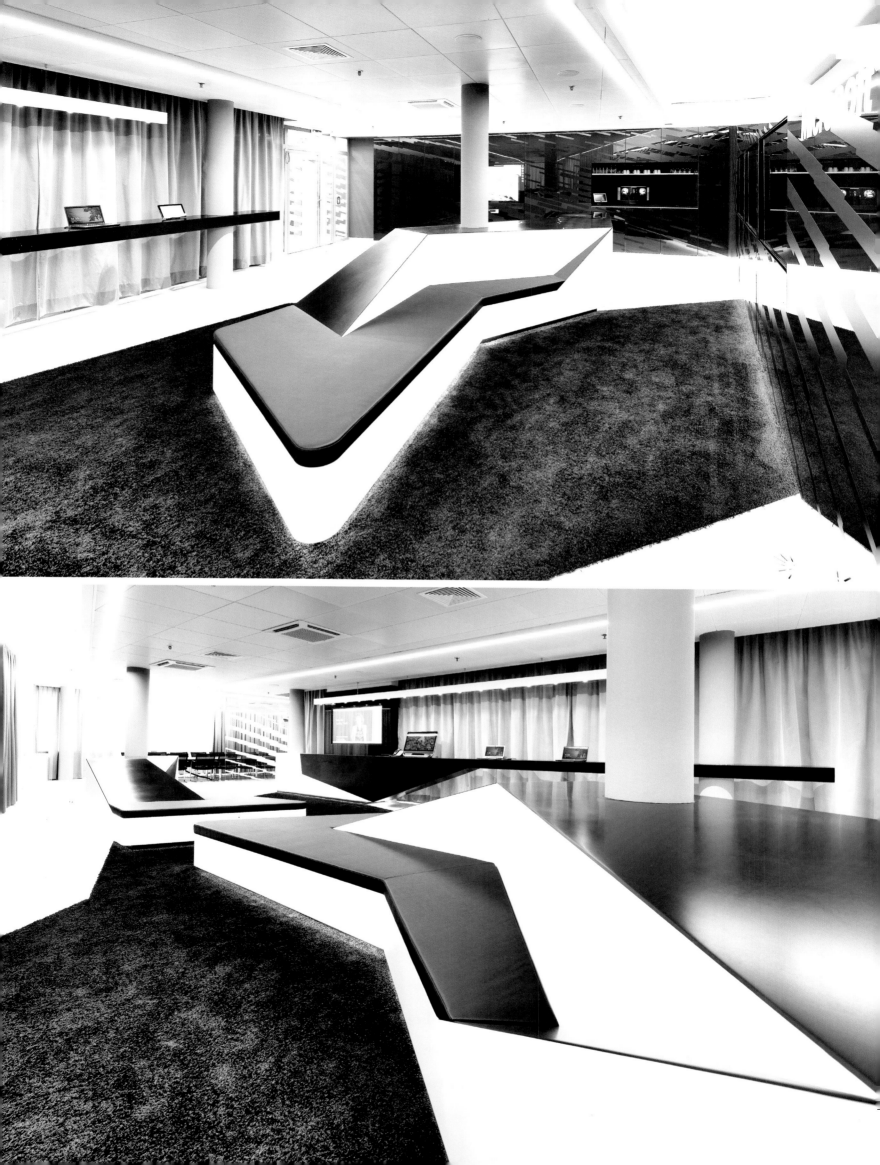

LOCATION Germany

ADIDAS WORKOUT

DESIGNER
KINZO

PHOTOGRAPHER
Werner Huthmacher

KINZO develops an office space and furniture program for adidas that combines highest functional needs with the spectacular laces-architecture by kadawittfeld. The furniture design absorbs the communicative architectural language and completes the office complex making it a dazzling experience with the visual quality and continuity of a film set – as if the furniture were part of the architecture or the architecture were part of the furniture.

Already in spring 2009, KINZO is able to win the idea competition competing against renowned manufacturers. The nifty and modular furniture concept accomplishes the required balance between various functional aspects virtually without effort: Large amounts of apparel and footwear samples can most naturally be combined with ordinary office articles.

The assortment of close to 50 elements enables numerous room and function variations. So the system presents itself as a well-casted and precisely coordinated furniture team with a sporting spirit. In the end, a new method of environment-friendly powder coating guarantees extra quality adding a premium finish to the progressively debonair design.

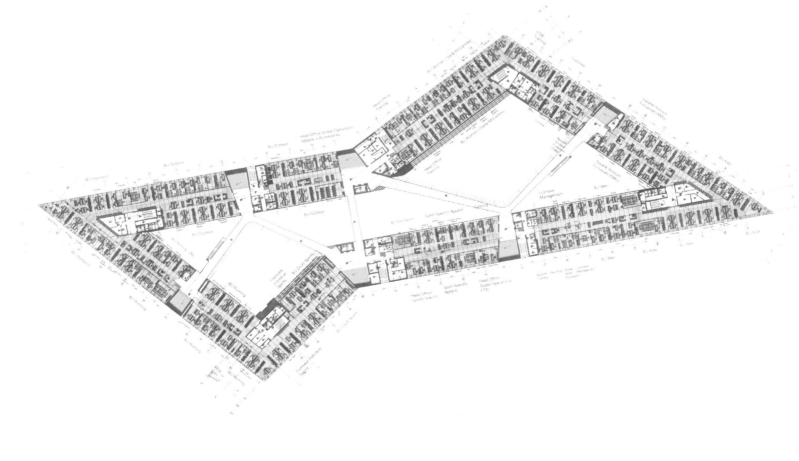

223

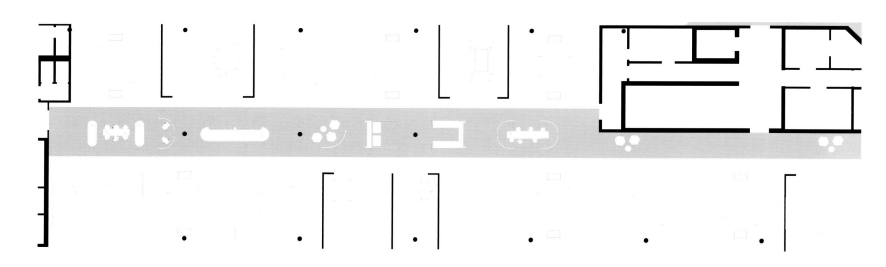

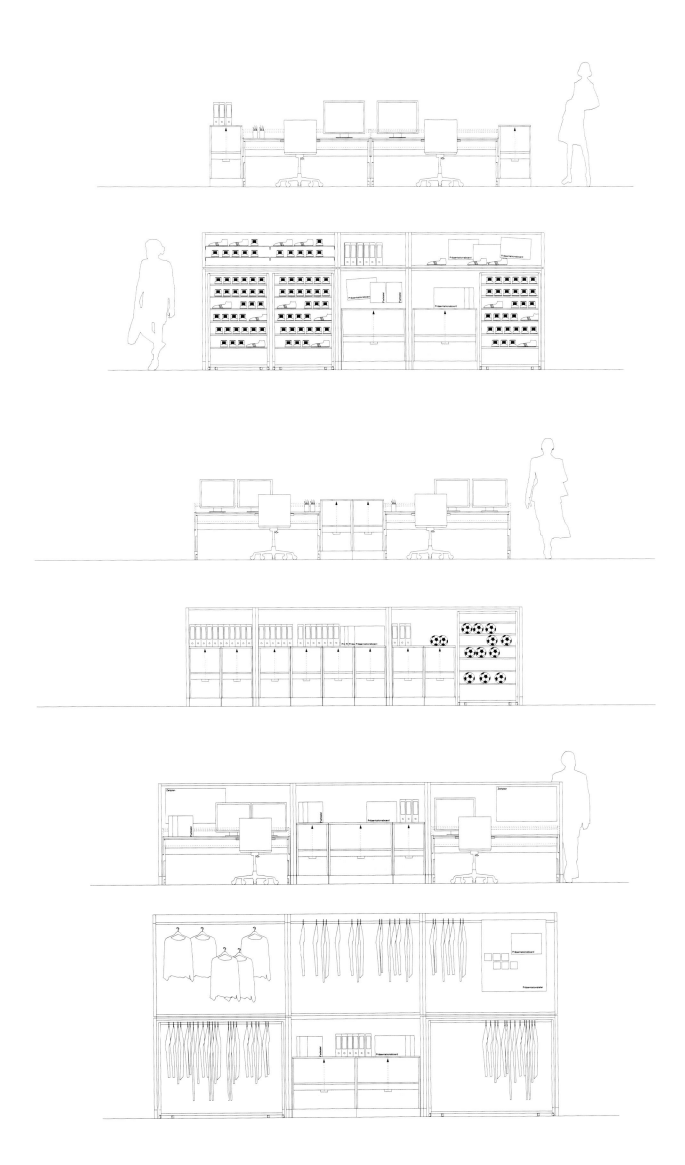

229

LOCATION Tokyo, Japan

GOOGLE TOKYO

DESIGNER
Klein Dytham architecture

PHOTOGRAPHER
Daici Ano, Koichi Torimura

Klein Dytham architecture (KDa) recently completed Google's new Japan office. This ambitious interior project is located in the Roppongi Hills tower in central Tokyo.

Reflecting Google's ambition to invent amazing things, KDa sought to create a fun, inspiring workplace with places for brainstorming, casual chats, or simply to hang out between bursts of work. These include a series of spaces each with a unique theme – a "hairy" blue café space clad in the huge brushes used in carwashes, and a presentation space created in the image of a traditional Japanese neighbourhood bathhouse.

Color and cultural imagery is used to help with wayfinding. KDa divided the interior into various zones and employed vivid textures and bold graphics to give each area a distinct character. Circulation routes were defined by block walls to suggest the winding lanes of Tokyo's residential areas, wall graphics were designed to recall Japan's timber architecture, and wallpaper patterns cleverly incorporate Google icons.

231

235

LOCATION Vienna, Austria

MICROSOFT HEADQUARTERS IN VIENNA

DESIGNER
INNOCAD

AREA
4,500 m²

PHOTOGRAPHER
Paul Ott, Christian Dusek

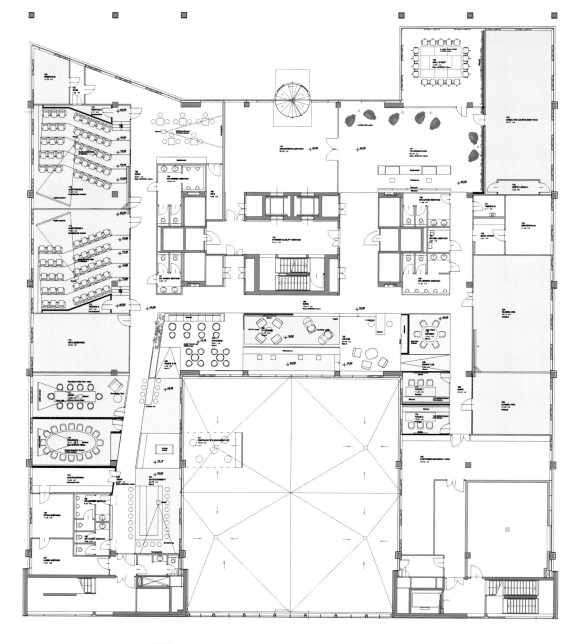

DATA HIGHWAY   ICF - INTERACTION CIRCULATION FURNITURE   MEETING ROOMS

In order to create the ideal constellation of physical, social and virtual work environments for the company's employees, Microsoft has conducted several studies on this topic in recent years.

The complete renovation of the 4,500m² headquarters in Vienna took this concept to heart and even took it one step further: the "sealed-off" employee floors were broken up and arranged in a transparent manner. An average of 10 percent greater employee satisfaction, a 12 percent increase in productivity and efficiency, improved CO_2 footprints and many other similar benefits can be expected – and some can even already be measured.

An architectural "life-line" traverses the entire building in the form of accessible, multi-functional furniture, providing a spatial bracket around all of the floors, and facilitates a variety of functional settings. The greatest possible flexibility is also provided in the closed meeting rooms: Every employee can select their ideal environment, according to their needs and mood. All high-traffic areas, such as corridors and foyers, were designed to be intentionally dynamic. The striped vinyl floor and a slide that allows quick access from the second to the first floor both symbolize movement. The accompanying green walls in all of the floors have both an atmospheric and positive effect on the climate of the space. The lighting concept works with as few light sources as possible. Neutral, uniform lighting in the form of linear light elements in the "life-line" creates an atmospheric undertone.

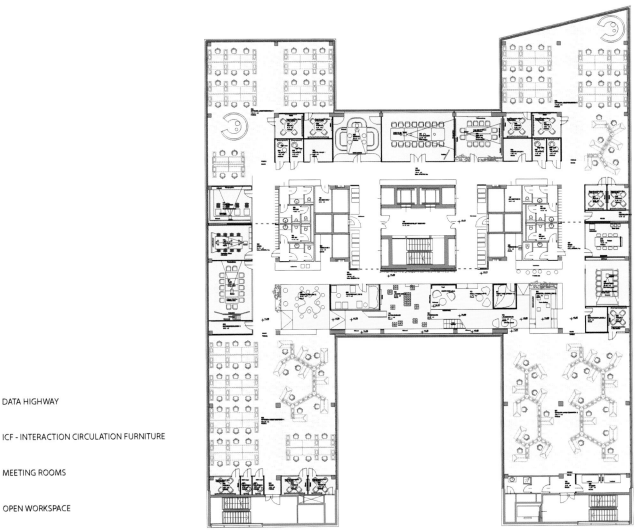

DATA HIGHWAY

ICF - INTERACTION CIRCULATION FURNITURE

MEETING ROOMS

OPEN WORKSPACE

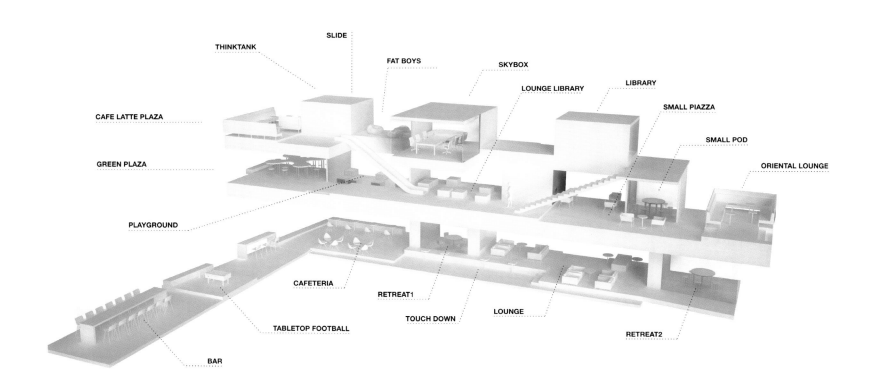

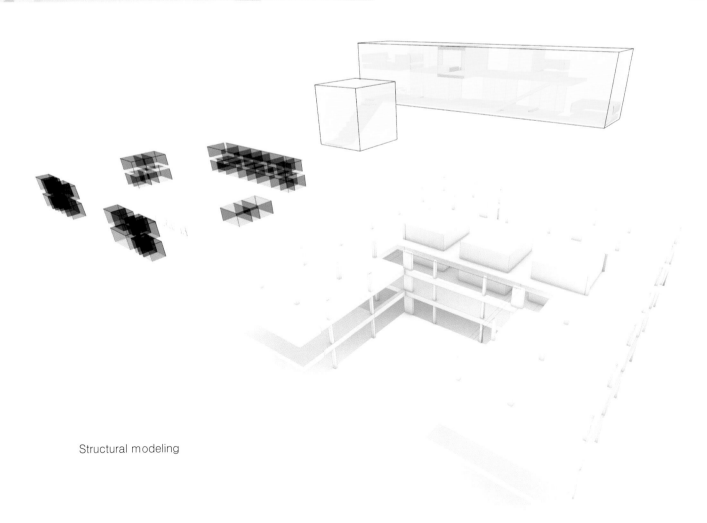

Structural modeling

LOCATION Toronto, Canada

RED BULL CANADA OFFICE

DESIGNER
Johnson Chou Inc.

PHOTOGRAPHER
Tom Arban

The client's brief was essentially three-fold: to create a space consistent in concept and form to the original yet be a reinterpretation of it; that the space be inspiring for staff in administrative/accounts who are normally relegated to bland, impersonal spaces; and that the predominant materials be that from a reclaimed source. Continuing the narrative or architectural concept of vessels for transformation, the design focuses on four main components: a bar/lounge area, a linear open workstation concept, a wool felt-clad form and a boardroom.

As Red Bull is host to many social gatherings, from art openings, parties and receptions, the bar/lounge area functions to anchor the public areas of the space. An aluminum mesh screen divides the bar from the open workstations. For public events the screens, suspended from tracks, slide to either wall to prevent entry. The custom workstations, composed like a ribbon, transform from a floor surface to a work surface to a ceiling plane, providing color, texture, warmth and a sense of enclosure. A defunct elevator core is re-formed and clad with varying colors and thickness of felt, utilized for its sound absorption qualities. The organic form is evocative of the natural contours of a mountain or the muscular torso of a bull.

A new meeting room composed of felt, blackboard and tamarack wood strips positioned in a random pattern is another vessel or "primitive hut". The lighting and the wood strips in the ceiling are intending to evoke the effect of light filtering through branches. A long boardroom table fabricated from a cherry tree felled on the millworker's own property anchors the space.

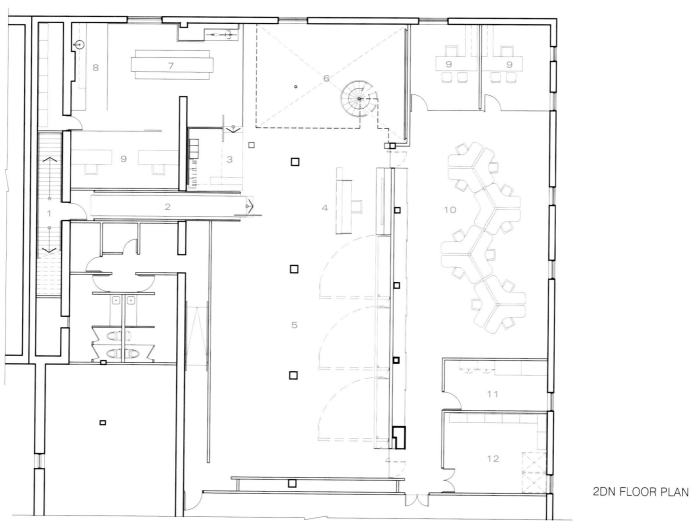

2DN FLOOR PLAN

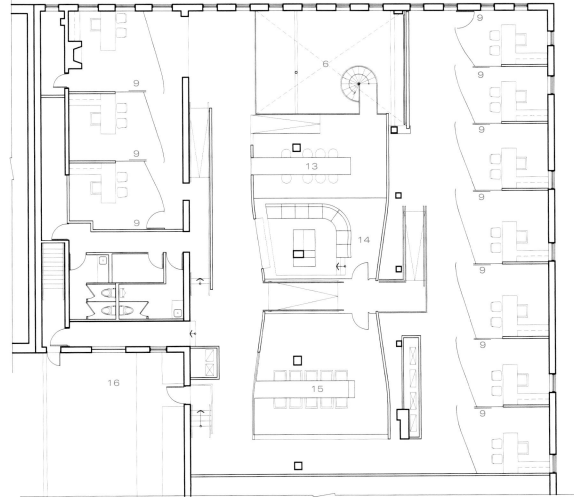

1. ENTRANCE STAIR
2. TRANSITION TUBE
3. WAITING / LIBRARY
4. RECEPTION
5. GALLERY
6. ATRIUM
7. DINING/ LUNCH AREA
8. KITCHEN
9. PRIVATE OFFICES
10. OPEN WORKSTATION
11. TEAM OFFICE
12. PRINTING STORAGE
13. BAR MEETING AREA
14. LOUNGE MEETING AREA
15. BOARDROOM
16. EXTERIOR DECK

LOCATION Melbourne, Victoria, Australia

ISELECT

DESIGNER
V Arc

AREA
4,800 m²

V Arc have recently completed an unconventional fitout for the iSelect team with a real wow factor, partnering with FDC Building + Construction.

The 4800m² site, split across three levels, marries contemporary design practices and space planning with clear design drivers: design an environment that is fun, quirky and gives back to the staff, and create an environment that is professional yet entertaining – promoting social interaction throughout the entire fitout.

iSelect embraced "out there" design solutions and really supported out of the box thinking. In that respect iSelect and V Arc were a perfect match. The Front of House makes use of the iSelect corporate colors, orange and white. With stark white epoxy floors and bold splashes of orange, the primary paint finish here is white, white and white, used to complement iSelect's palette.

Further throughout the fitout, V Arc employed an array of colors, grounded with iSelect orange, white and black. Pops of color were integrated into all aspects of the fitout, maintaining a lively atmosphere and providing visual relief for staff. Informal touchdown spaces are punctuated with feature lighting and seating, once again reinforcing the iSelect brand. Feature paint finishes were employed here to complement and contrast with the commercial palette.

While the "watering holes" pick up on the iSelect orange, they are carved from the negative space around the stairwell, highlighted in Dulux Black to contrast from the front of house stark white finish.

From the bright green freestanding meeting pods to the blue of the indoor soccer pitch to the black tones of the breakout and 300m² cafe, color is used as a visual indicator and reference point around the floor.

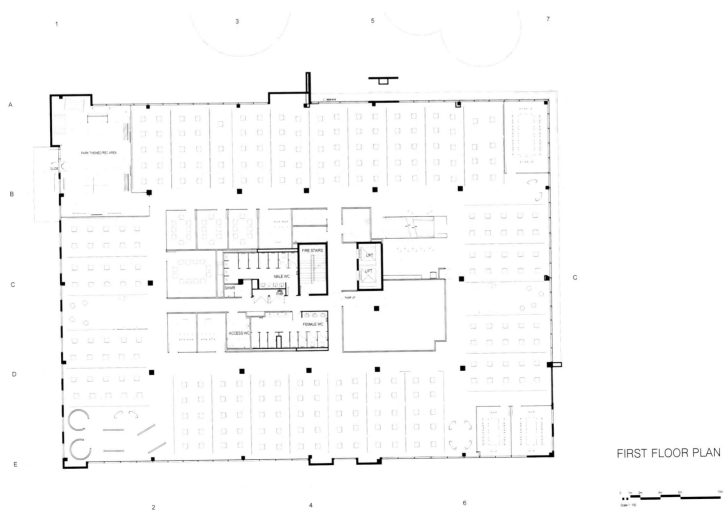

FIRST FLOOR PLAN

GROUND FLOOR PLAN

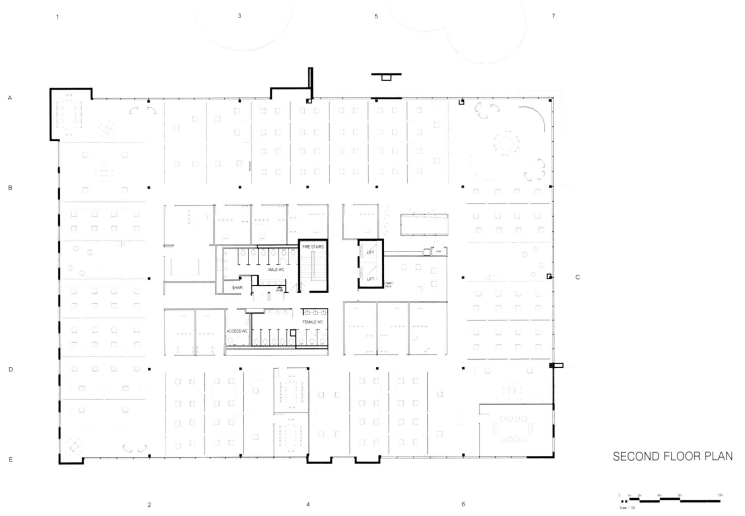

SECOND FLOOR PLAN

LOCATION Kazan, Russia

KAZAN YANDEX OFFICE

DESIGNER
za bor architects

AREA
674 m²

PHOTOGRAPHER
Peter Zaytsev

Representative office of the largest Russian IT-company, Yandex in Kazan, is situated on the 16th floor of Suvar Plaza business center. The office plan reminds a trapezium, while the midline of a trapezium is a corridor, leading from the main entrance to the fire exit.

The office volumes are organized in a corridor system – both side of the corridor has 4 open space blocks, designed for 6~13 working places. There is also a separate server zone, which borders with a storage and IT-administrator's office. Front-office volumes are located close to the main entrance – there is a reception desk, two meeting rooms, lecture-hall and canteen. Next there are working places and the chief's office, which looks like a glass polygonal lamp that functions as a zoning element, partially separating two nearest open spaces. If it is necessary, the chief's office may be isolated with curtains. Beyond that, matted glass panels are used as design element. At the end of the corridor there are gym and shower area.

To make the working areas more comfortable, the lighting estimations have been made, to correctly organize lighting, and the ceilings have been sheeted with Ecophon acoustic panels, which have sound-absorbing qualities.

Traditionally Yandex offices are famous for their informal rooms – unique semi-closed "cells", isolated from the corridor volume. In this office, the architects used rounded shapes and industrial carpet, which covers thoroughly the inside area of the two "cells". Outside they are white, as well as lines of communication up across the ceiling, which gives an effective contrast to black wooden floors. The only intensive colors in the office, different from neutral monochrome colors, is a contrasting color scheme of terracotta-green, which was used repeatedly in textile and furnishings.

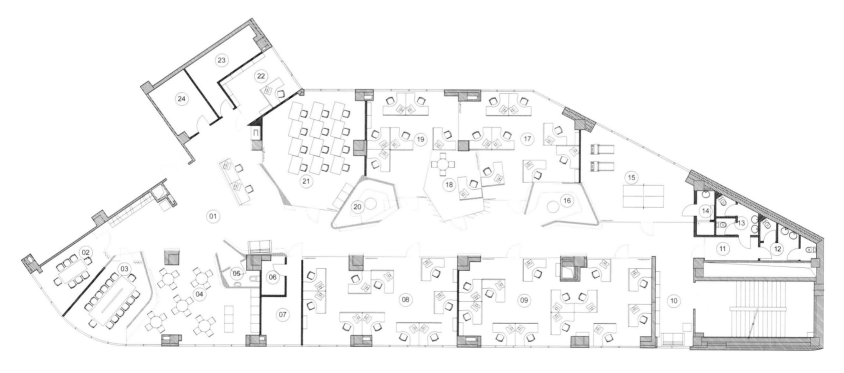

LOCATION San Francisco, California, USA

SQUARETRADE

DESIGNER
Design Blitz

AREA
4,924 m²

PHOTOGRAPHER
Bruce Damonte

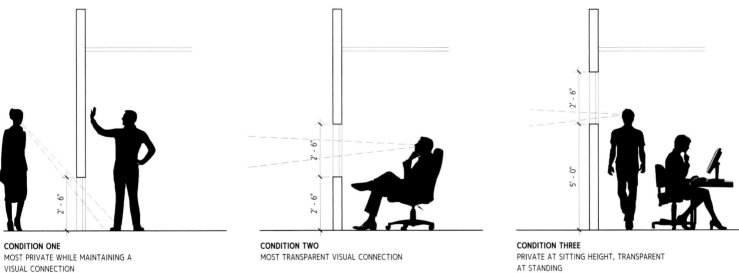

CONDITION ONE
MOST PRIVATE WHILE MAINTAINING A
VISUAL CONNECTION

CONDITION TWO
MOST TRANSPARENT VISUAL CONNECTION

CONDITION THREE
PRIVATE AT SITTING HEIGHT, TRANSPARENT
AT STANDING

Before SquareTrade moved into their new San Francisco office in a mixed-use building on Third Street, the interior had been outdated and recalled a bygone era of tech.

SquareTrade hired Design Blitz architects to envision the new space, all within a tight budget and on an incredibly short schedule. Blitz democratized the office layout by organizing workstation seating along the glazed exterior walls so that all could partake in the natural light and fantastic views. The open office layout created a dilemma though, as the heart of the office is a large, central cafe area that accommodates not only dining functions, but also meetings and presentations, and therefore produces a great deal of sound. To mitigate the acoustics, Design Blitz created two large banks of conference rooms at a 20 degree angle off-grid that run parallel to the hive. The bars not only buffer the office area from the cafe, but also add visual appeal to the space, as their placement achieves unique sight-lines and divides remaining open spaces into additional breakout areas.

To give the bars an architectural, three-dimensional quality, Blitz applied walls as a "skin" element to them. The skins extend several inches from the edges of the conference rooms, creating a floating, subtly futuristic effect. Offering visual appeal in addition to functionality, the conference rooms have glazed slits at varying heights according to each room's purpose (ground level slits for high privacy, desk level for medium privacy, and standing height for maximum privacy), creating an architectural permeability that corresponds with the function of each room.

The overall design of the project was inspired by the rectilinear nature of SquareTrade's logo, with an emphasis on clean, white edges.

LOCATION San Francisco, California, USA

ATLASSIAN OFFICES

DESIGNER	DESIGN COMPANY	DESIGN TEAM MEMBER	AREA	PHOTOGRAPHER
Sarah E. Willmer	Studio Sarah Willmer Architecture	Doris Guerrero, Megan Carter, Josue Munoz-Miramon, Olya Piskun	3,902 m²	Jasper Sanidad

The new headquarters for Atlassian Inc is a 3,902m² adaptive re-use of a freestanding 1924 industrial building on Harrison Street in San Francisco's SoMa district. The location offers proximity to neighborhood amenities and mass transit with easy bike access for employees and clients. Generous bike storage, office share bicycles, and employee showers are provided.

The building was selected by the architect-client team for its urban location, passive environmental features, and long-term occupancy. The building offers excellent natural ventilation, daylight to all employee work areas, and ample, flexible collaborative workspace for employees and clients.

Demolition was selective to maximize re-use and to create ease of circulation. The existing base building, structure, exterior glazing, elevator, bathrooms, exit stairs, mechanical rooms and server rooms remained untouched. Some existing rooms remain and many salvage doors have been reconditioned and reinstalled. Rather than re-carpeting the entire building, newly exposed concrete floors reveal a natural finish surface at major public spaces. Existing floor tile and accessories are re-used in the restrooms. 80% of the existing lighting fixtures remain in place or have been relocated.

An existing clerestory provides natural daylight to all employee work areas. New conference rooms are glazed, taking advantage of the daylight. Additionally, the clerestory is operable and provides fresh air and natural ventilation throughout the space. Windows on the west and south walls utilize solar shades to mitigate heat gain and glare at workstations. The existing mechanical system has been re-used and augmented only where necessary at primary conference rooms.

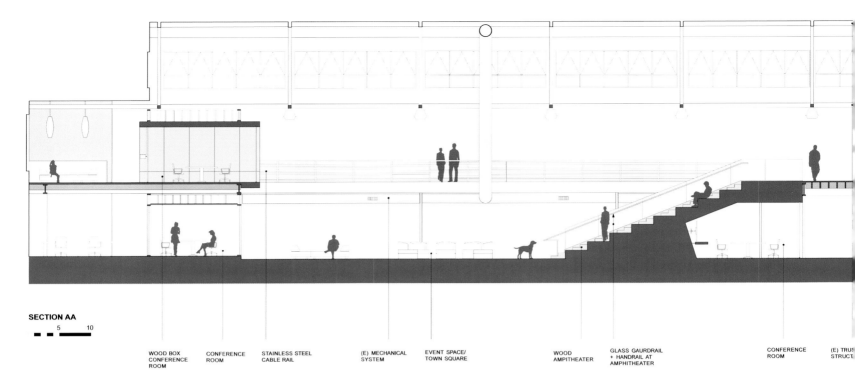

SECTION AA

Atlassian II
Offices

WOOD BOX CONFERENCE ROOM | CONFERENCE ROOM | STAINLESS STEEL CABLE RAIL | (E) MECHANICAL SYSTEM | EVENT SPACE/ TOWN SQUARE | WOOD AMPITHEATER | GLASS GAURDRAIL + HANDRAIL AT AMPHITHEATER | CONFERENCE ROOM | (E) TRUSS STRUCTURE

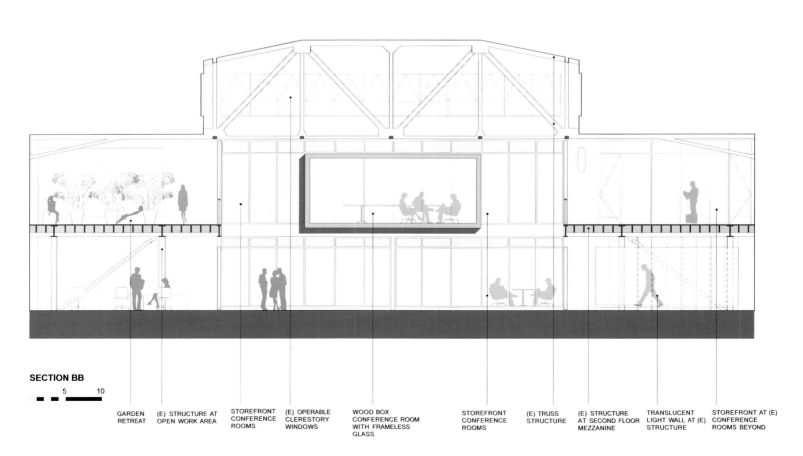

SECTION BB

Atlassian II
Offices

GARDEN RETREAT | (E) STRUCTURE AT OPEN WORK AREA | STOREFRONT CONFERENCE ROOMS | (E) OPERABLE CLERESTORY WINDOWS | WOOD BOX CONFERENCE ROOM WITH FRAMELESS GLASS | STOREFRONT CONFERENCE ROOMS | (E) TRUSS STRUCTURE | (E) STRUCTURE AT SECOND FLOOR MEZZANINE | TRANSLUCENT LIGHT WALL AT (E) STRUCTURE | STOREFRONT AT (E) CONFERENCE ROOMS BEYOND

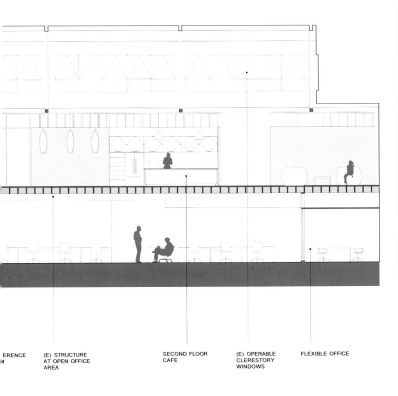

| ERENCE | (E) STRUCTURE AT OPEN OFFICE AREA | SECOND FLOOR CAFE | (E) OPERABLE CLERESTORY WINDOWS | FLEXIBLE OFFICE |

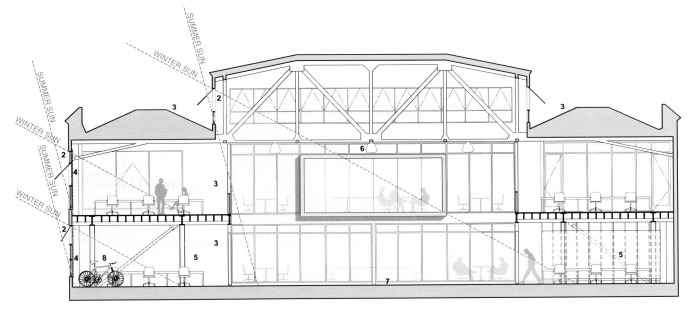

SUSTAINABLE STRATEGIES

LEGEND:

1. **ADAPTIVE REUSE**
   (E) BASE BUILDING: ELEVATOR, RESTROOMS, EXIT STAIRS, MECHANICAL ROOMS, SERVER ROOMS, STRUCTURE AND EXTERIOR GLAZING REMAIN

2. **NATURAL DAYLIGHT / PASSIVE HEAT**
   (E) WINDOWS PROVIDE NATURAL LIGHT TO ALL EMPLOYEE WORK AREAS AND PASSIVE SOLAR HEATING

3. **NATURAL VENTILATION**
   (E) OPERABLE WINDOWS AND CLERESTORY PROVIDE FRESH AIR

4. **SOLAR SHADING / GLARE CONTROL**
   MITIGATES HEAT GAIN AND SCREEN GLARE TO WORKSTATIONS

5. **AMBIENT LIGHT**
   (E) WINDOWS AND CLERESTORY PROVIDE NATURAL AMBIENT LIGHT THROUGHOUT

6. **LIGHTING**
   80% OF (E) LIGHTING FIXTURES WERE RE-USED AND/OR RELOCATED

7. **CONCRETE FLOOR**
   EXPOSED CONCRETE FLOOR PROVIDES THERMAL MASS FOR SUMMER COOLING

8. **ALTERNATIVE TRANSPORTATION**
   AMPLE BIKE STORAGE PROVIDED FOR EMPLOYEE BIKES; ADDITIONAL BIKES AVAILABLE FOR LOCAL USE

9. **PUBLIC TRANSPORTATION**
   5 BUS ROUTES WITHIN SEVERAL BLOCKS; LOCATED LESS THAN A MILE FROM CALTRAIN AND BART

Atlassian Offices

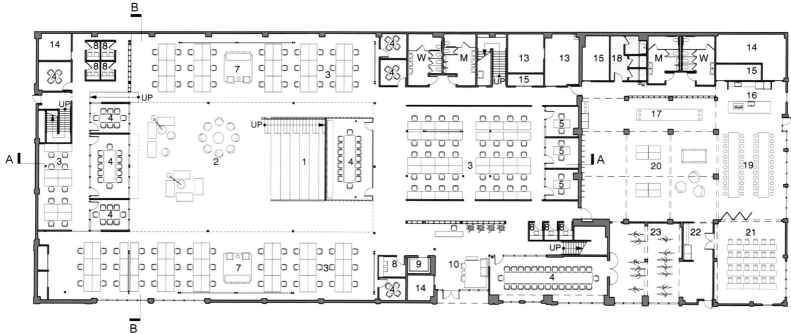

FIRST FLOOR PLAN

0 5 10   20   30

Atlassian Offices

1. AMPHITHEATER
2. EVENT SPACE/ TOWN SQUARE
3. OPEN WORK AREA
4. CONFERENCE ROOM
5. FLEXIBLE OFFICE
6. COMMON LOUNGE
7. TEAM LOUNGE AREA/ FUTURE WORKSTATIONS
8. SKYPE BOOTH
9. ELEVATOR
10. ENTRY
11. GARDEN RETREAT

12. 2ND FLOOR CAFÉ
13. SERVER/ I.T. ROOM
14. MECHANICAL ROOM
15. STORAGE ROOM
16. MAIN KITCHEN
17. SANDWICH ISLAND
18. COED. SHOWERS
19. DINING AREA
20. GAME ROOM
21. TRAINING ROOM
22. COFFEE BAR
23. BIKE STORAGE

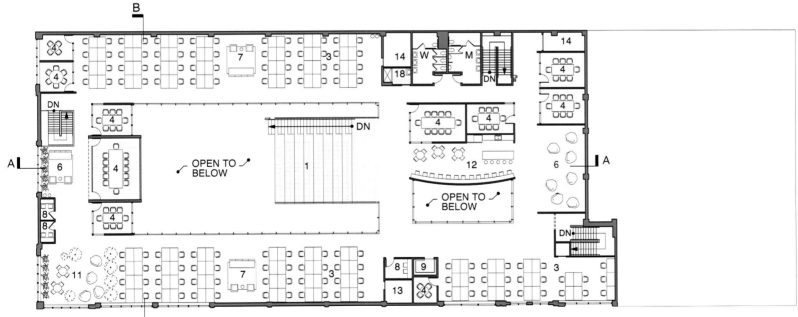

SECOND FLOOR PLAN

Atlassian Offices

1. AMPHITHEATER
2. EVENT SPACE/ TOWN SQUARE
3. OPEN WORK AREA
4. CONFERENCE ROOM
5. FLEXIBLE OFFICE
6. COMMON LOUNGE
7. TEAM LOUNGE AREA/ FUTURE WORKSTATIONS
8. SKYPE BOOTH
9. ELEVATOR
10. ENTRY
11. GARDEN RETREAT
12. 2ND FLOOR CAFÉ
13. SERVER/ I.T. ROOM
14. MECHANICAL ROOM
15. STORAGE ROOM
16. MAIN KITCHEN
17. SANDWICH ISLAND
18. COED. SHOWERS
19. DINING AREA
20. GAME ROOM
21. TRAINING ROOM
22. COFFEE BAR
23. BIKE STORAGE

LOCATION Bostancı, Istanbul, Turkey

SAHIBINDEN.COM OFFICE

DESIGNER
Erginoğlu & Çalışlar Architects

5,000 m²

PHOTOGRAPHER
Emre Dörter

The Sahibinden.com project is an office in the Bostancı district at the Asian side of Istanbul, with a total area of 5,000 m² on two floors. The office has been designed in accordance with the spirit and corporate stance of this technology company, to enhance employee productivity and overall comfort.

The board room, meeting rooms and most of the working areas were located on the second floor, the plan of which was guided by the elliptical form of the building. The working area, which for the most part was designed as an open-plan office, and closed spaces like the meeting and board rooms were resolved without relation to the facade, again on account of the building's shape. When planning spaces belonging to the meeting and board rooms inside the entrance to the second floor, amorphic forms were created by means of blind walls as a means of making best use of the space and in the attempt to disconnect them from the office areas.

A running track and the idea of integrating the recreational functions with the working areas was given precedence. Another important aspect of the planning of this floor is the meeting area located right at the center of the building. This space is reinforced by entertainment areas, such as a games room and TV corner, where the employees can spend free time, and which can double as spaces for different types of meetings. Off this section are a basketball court, fitness and cafe areas and an office. The latter space is the "brainstorming" area, which the employees of sahibinden.com, having a young and dynamic structure, can use for developing new ideas. For this they needed a conducive space apart from the office areas, and it has been designed so that they feel they are in a different atmosphere.

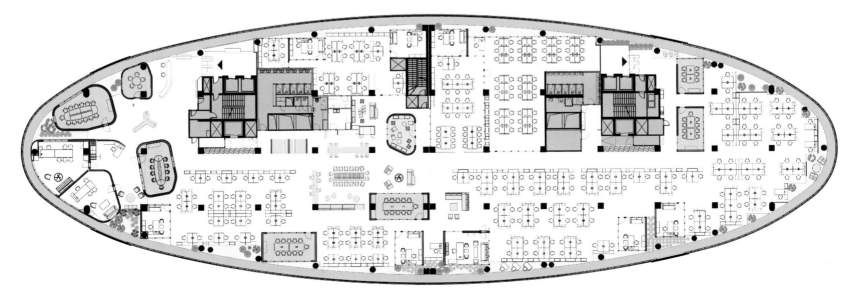

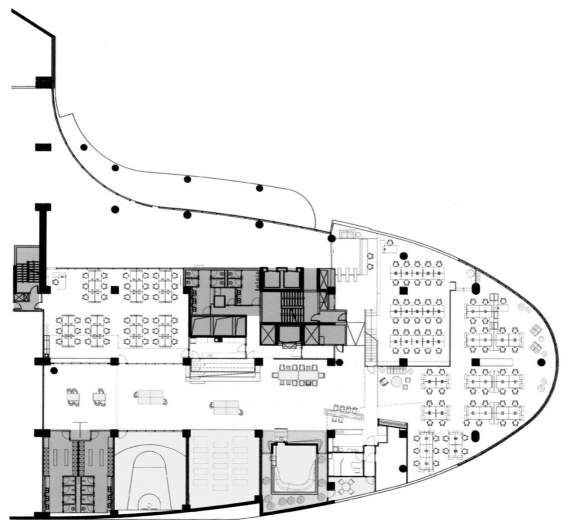

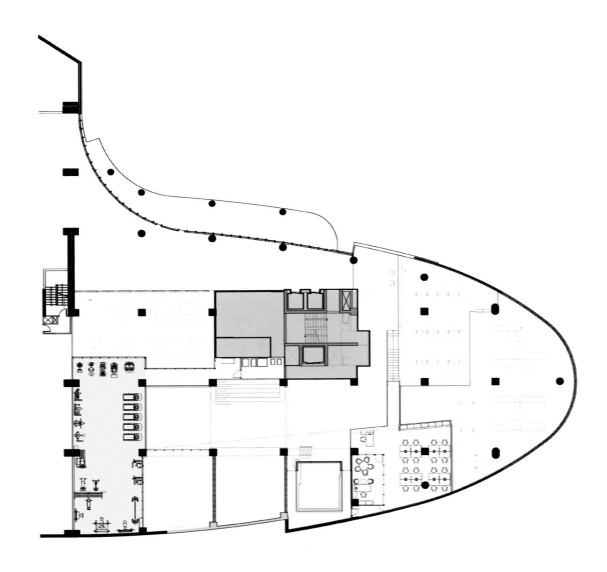

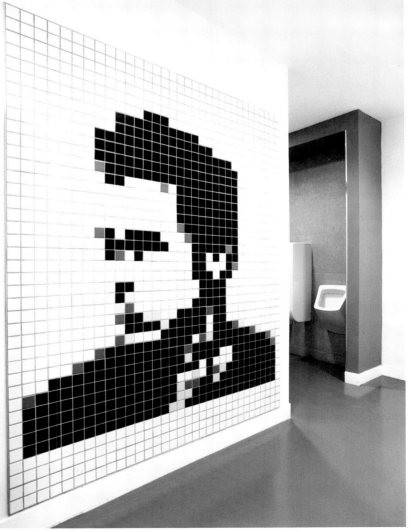

— **LOCATION** Balmumcu, Istanbul, Turkey —

YEMEKSEPETI.COM OFFICE

DESIGNER
Erginoğlu & Çalışlar Architects

AREA
2,000 m²

Sprawled across 2,000 square meters over five full floors and a basement/parking floor, the existing six-storey building was reorganized and converted into an office space for the Yemeksepeti project.

The primary objective of the project was to reflect the young, dynamic, unique, and fun working environment of the company on the office spaces. With this idea in mind, the project foresaw well-lit and spacious areas that would arouse a feeling of belonging among the employees.

Colorful and vivid working paces with a sense of character were thus envisaged. In addition, recreation, meeting, and play areas – including PlayStation, table tennis, basketball, etc. – were designed to allow the predominantly young employees to take a break from the daily stress of the work environment.

The focal point of the project was the transformation of the existing car elevator between the ground and parking floors into meeting room in motion. Hence, the meeting room was designed to travel to the floor on which the meeting would be held such that visitors could observe the building and the work environment of the office from a different perspective.

The top floor of the office building is designed as an activity and meeting space. Due to the furniture comprised of movable modules and stands that can be opened or closed at will, this floor makes room for various group activities and screenings.

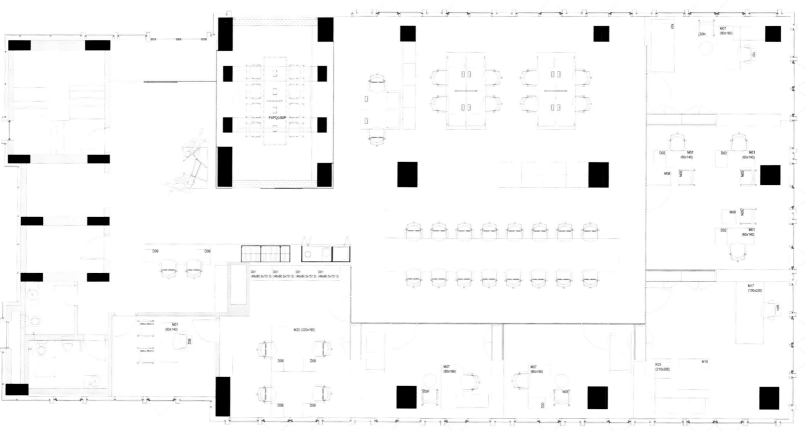

ground floor

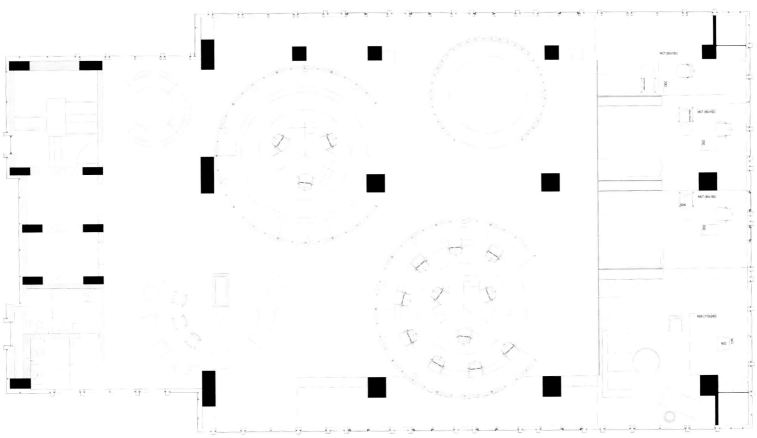

1st floor

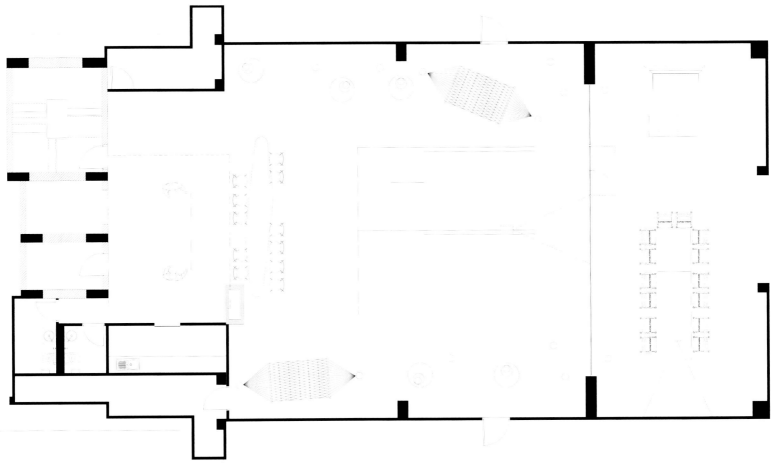

2nd floor

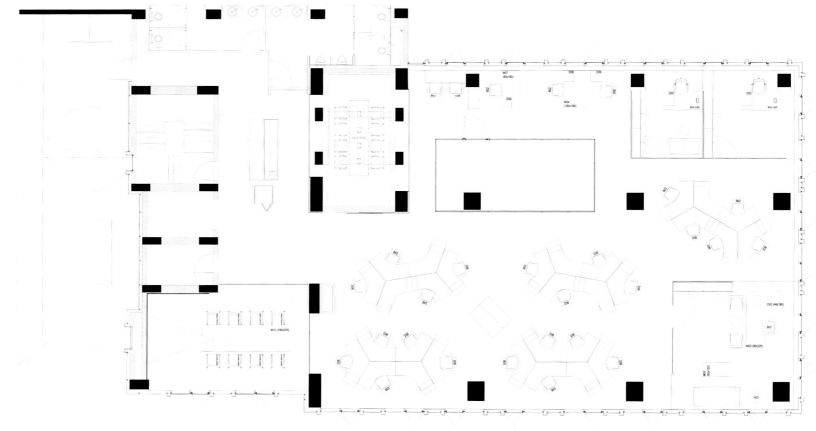

basement 1

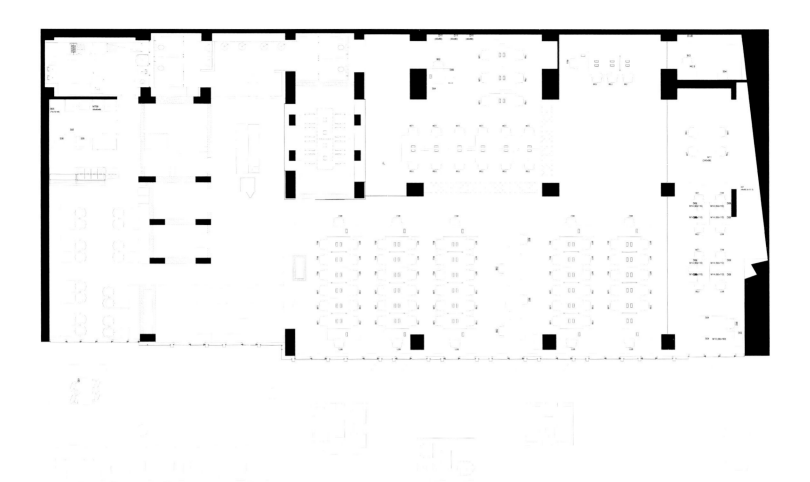

basement 2

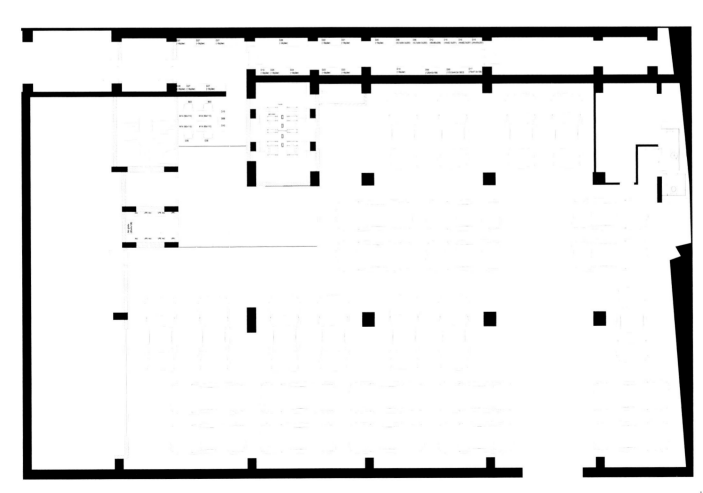

basement 3

LOCATION Moscow, Russia
OFFICE+SHOWROOM FOR DK

DESIGNER
Megabudka

Objective: To design bright multifunctional space convenient for office work, visual for demonstration of interior elements, comfortable for lectures and prepossessing to communication at friendly meetings.

Concept: We departed from regular view on interior by getting free from complexes of horizontal floor by floor project and placed complicated art object consisting of parallelepipeds into original empty space.

Internal color of each parallelepiped depends on its function. You can intricately move between the functions. And the object itself zones initial space around.

Now new office+showroom for DK Project company is a bright point of design and communications in Artplay, it unites designers, architects and creative people who appreciate quality and are ready to share new ideas, collaborate and develop concepts of modern design.

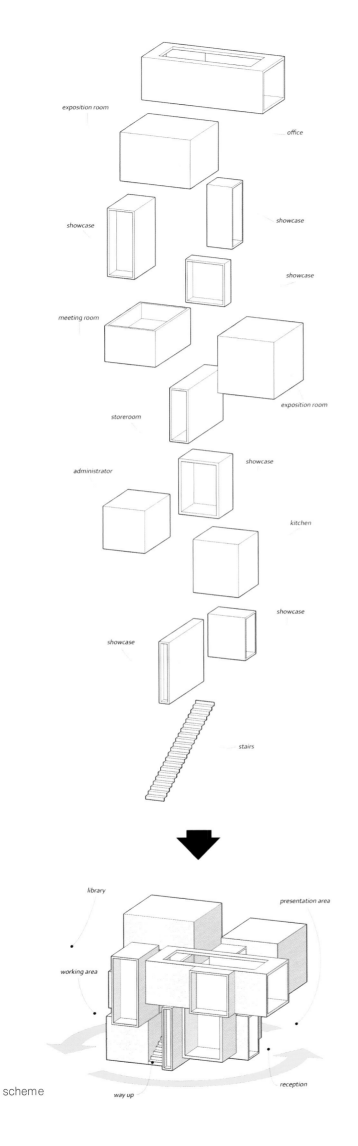

scheme

LOCATION Amsterdam, the Netherlands
AEGIS MEDIA

DESIGNER
D+Z Architecten+Projectmanagers

AREA
4,500 m²

Aegis Media rented the first four floors in the office building Ydek in Amsterdam. Aegis Media asked D+Z to design and realize the new office interior for their 200 employees. Aegis Media is a parent company of several international brands, who all have their own "section" in the new office. D+Z was already involved with the project during the development stage of the building. This gave Jeffrey van der Wees a perfect opportunity to include some features (for example raised floor in combination with concrete core activation) to take full advantage of the office space. The raised floor was necessary in order to create a pavilion in the interior. With this pavilion the "human measure" came back into the building and it gave a perfect opportunity to centralize the supporting functions (meeting places, pantry, bar, concentration rooms, copiers, etc.).

The pavilion is surrounded by a natural strip of open space workstations. The strip of open space workstations is interrupted by glass offices. Jeffrey purposely chose offices with glass walls to preserve the transparent environment and allow as much natural light as possible into the office. The same concept is repeated on every floor. Every brand received its own color on the pavilion and where necessary the brands are divided by glass walls.

The overall result is a sustainable interior where the playful design is in perfect balance with the robust image of the building.

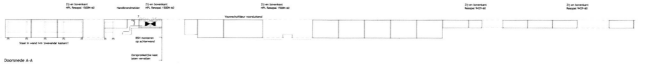

Cupboard wall

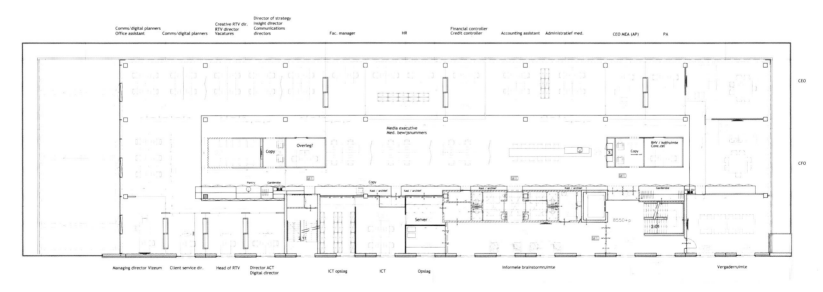

Layout

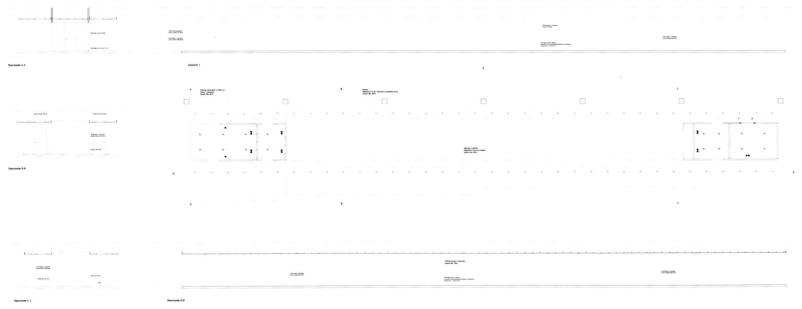

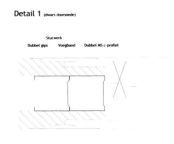
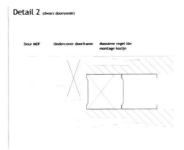

Pavilion

LOCATION Huntington Beach, California, USA

GRUPO GALLEGOS CREATIVE HEADQUARTERS

DESIGNER
Lorcan O'Herlihy Architects

AREA
2,787 m²

PHOTOGRAPHER
Lawrence Anderson(Esto)

The Grupo Gallegos Creative Headquarters is a non-traditional workplace that supports a Spanish language advertising company's unique culture. Located adjacent to the pier in Huntington Beach, California, USA, this project incorporates elements of the robust beach culture into its design.

The office's vibrant colors aesthetically engage with the adjacent coastal lifestyle, and a gymnasium, key to the initial programming of the space, caters to local youth sport leagues, bringing the community into the highly creative environment. LOHA utilized much of the existing structure as an opportunity. An amphitheater-stairway creates a transition between departments while providing a collaborative meeting and event space, the existing ticket booth was transformed into a juice bar, and mezzanine level conference rooms hover over the open floor plan below.

These protruding chambers double as billboards; when their louvered facades are pulled shut, the emblazoned word "ocupado" is formed to affirm the room's use. Over 300 white sunbrellas that allude to the reflective umbrellas of photography studios are repurposed to brighten the space and conceal fire protection. The resulting space resourcefully dials into a contextual economy of means.

The broad programming and refreshing design thus cultivate an environment that feels as much like a community hub as it does a workplace.

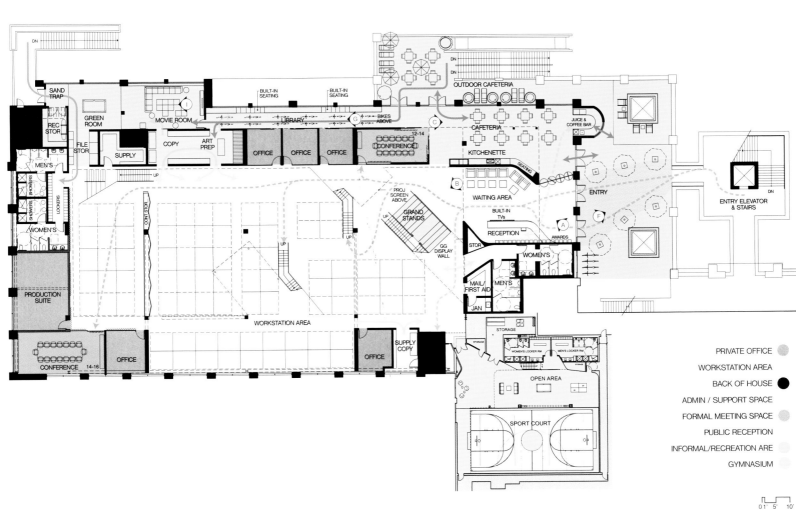

PRIVATE OFFICE
WORKSTATION AREA
BACK OF HOUSE
ADMIN / SUPPORT SPACE
FORMAL MEETING SPACE
PUBLIC RECEPTION
INFORMAL/RECREATION ARE
GYMNASIUM

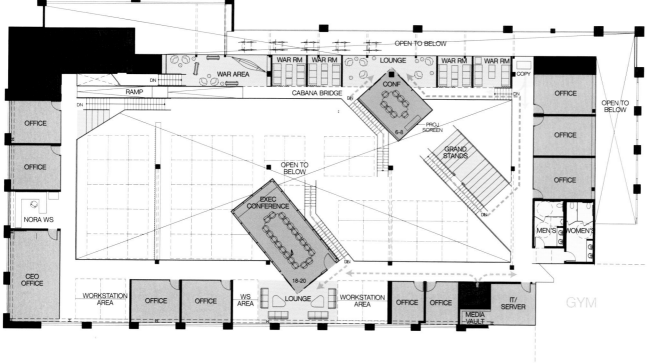

- PRIVATE OFFICE
- WORKSTATION AREA
- BACK OF HOUSE
- ADMIN / SUPPORT SPACE
- FORMAL MEETING SPACE
- PUBLIC RECEPTION
- INFORMAL/RECREATION AREA

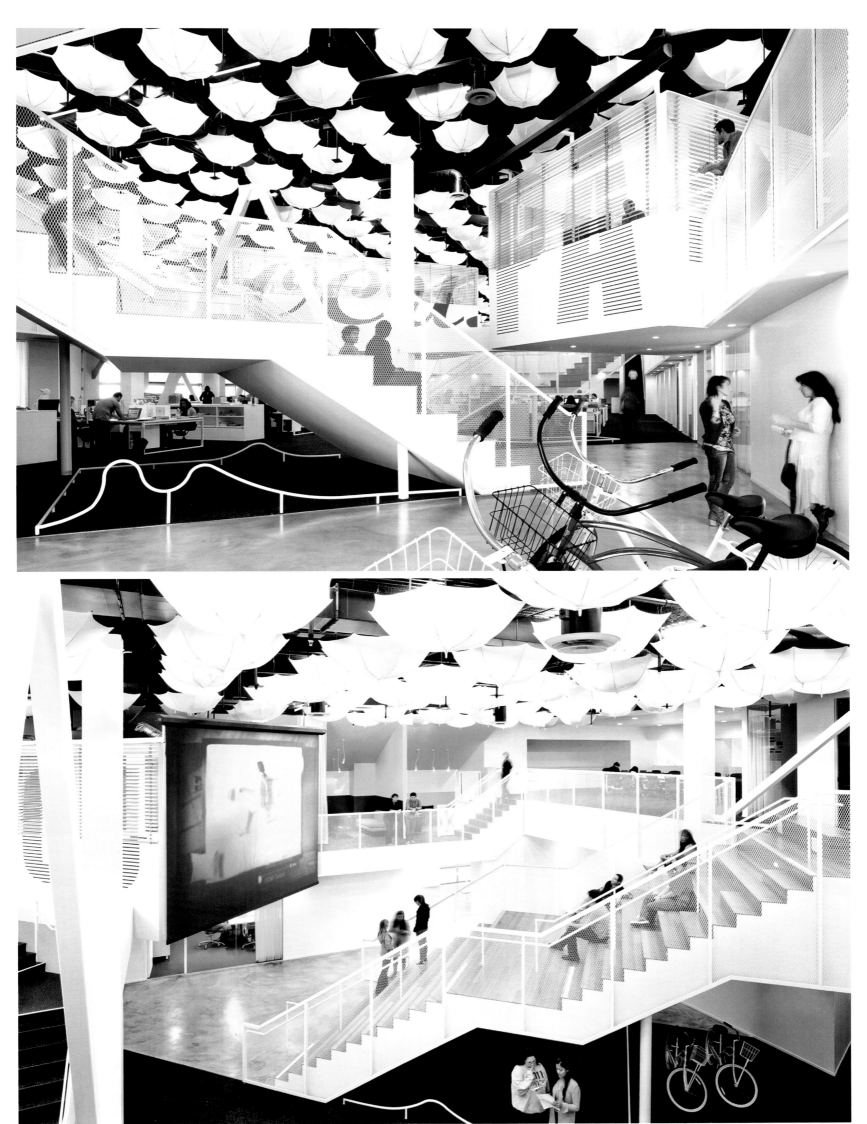

LOCATION Canada

ASTRAL MEDIA

DESIGNER
Lemay

AREA
6,000 m²

PHOTOGRAPHER
Mr. Claude-Simon Langlois

In the spring of 2010, Astral Media client relocated approximately 350 employees to four floors of a downtown Montreal building. The goals of this large-scale project included elaborating new furniture standards, fitting up flexible meeting spaces and optimizing employee interconnectivity in a contemporary, energetic and versatile working environment.

In order to create a rhythm and a gradation throughout the playful 6,000m² space, each floor is identified by its own color and the levels are linked by a central glass staircase.

The project's main objective consisted in rejuvenating the client's brand and accommodating a large number of previously dispersed employees in a single open and standardized space, while being mindful of generational differences in the staff makeup.

Graphic and signage interventions were incorporated into the project's design from the get-go. These elements are an integral and indivisible part of the space and animate the environment while giving it purpose, personality and functionality. The concept plays on the contrast between medium and message and manifests itself in undulating and pixelated graphics.

Aside from the ingenious use of graphics, light and color to express Astral's new identity, the project's creativity is also expressed through clever space usage. The result: a compact but ultra-functional working environment accommodating formal and informal meeting spaces such as cafes, meeting rooms, an auditorium and a game room, all linked by the central glass stairway, the spine of the entire project. And to what end? Incite more interaction between the different business units and promote impromptu exchanges between employees. In other words, communication!

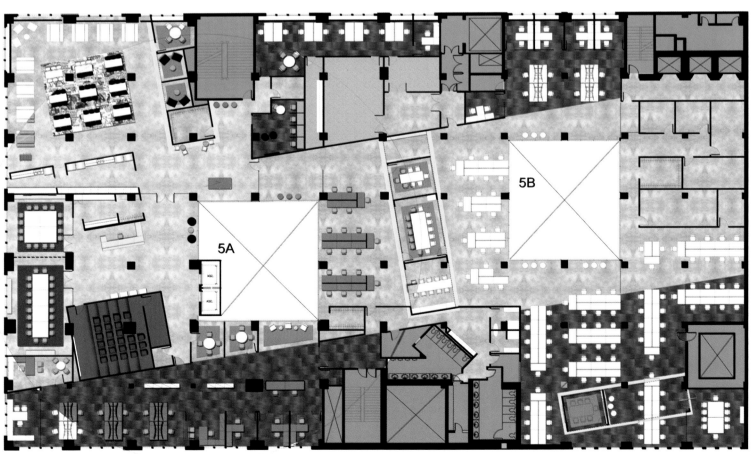

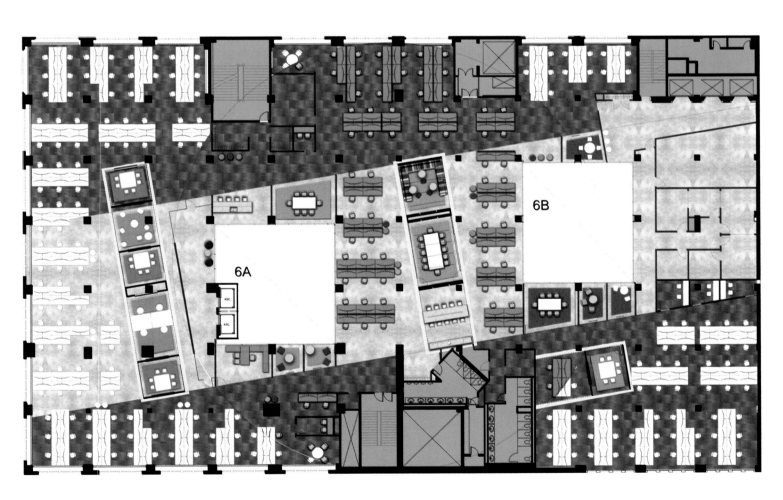

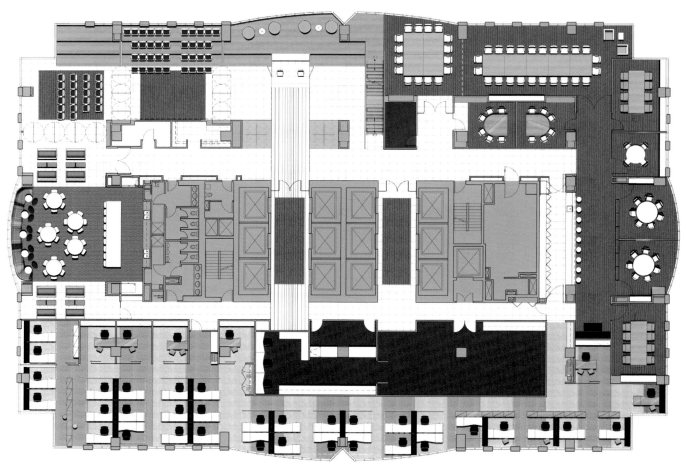

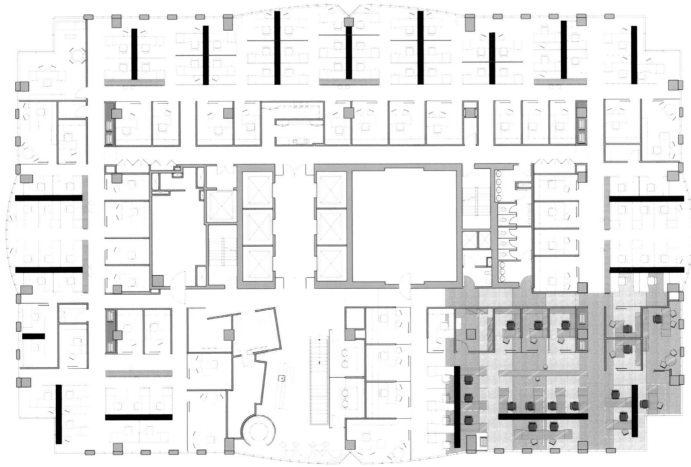

LOCATION Southampton, UK

PEER 1 OFFICE

Space & Solutions

In July 2012, Peer 1 Hosting announced plans to upscale and relocate their Southampton UK office to make it their EMEA Head Office.

The design brief to lead designer Sarah O'Callaghan, Managing Partner of Space & Solutions was to make it the best place in the UK to work – innovative and inspiring workspace that will attract the best talent. The client wanted people to walk in and think "yes I really want to work here".

Sarah commented, "Their business didn't require a traditional approach it wasn't about getting the most people possible into the space. This building works because it underpins their unique culture. They aren't fixed to desks, everybody has a laptop, they are constantly on wireless headsets, walking around and interacting. We collaborated closely with staff to tell us what they liked and what they didn't. We've created big, open, expansive spaces, with little pockets of interest throughout. Every space has been designed with a multifunctional edge to it.

Features include an indoor garden complete with a three meter high tree, a tree house, swings suspended from the ceiling and a mini golf course. There is a cinema room equipped with its own popcorn maker, a grass covered amphitheatre reception, an amazing 2,500 square feet($762m^2$) patio, a 'proper' English pub, a 'homely styled' canteen, coffee bar and themed lounge areas and two fully furnished flats for overseas guests. Not forgetting a giant adult slide for a quick exit."

Sarah concluded, "this was an amazing brief, a chance to do something different that would complement a unique and forward thinking company, where working hard and having fun at the heart of what they do."

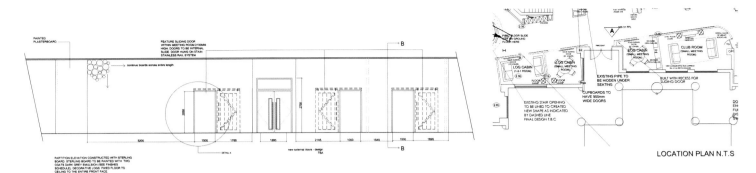
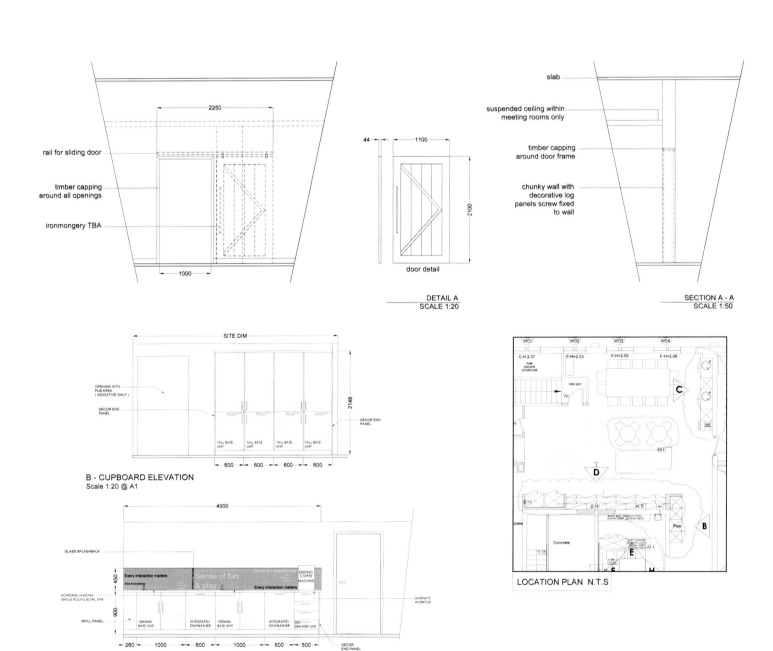
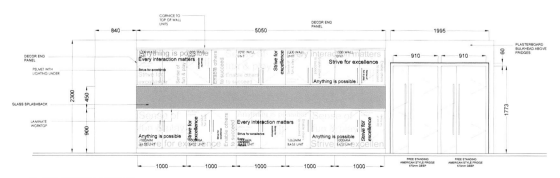

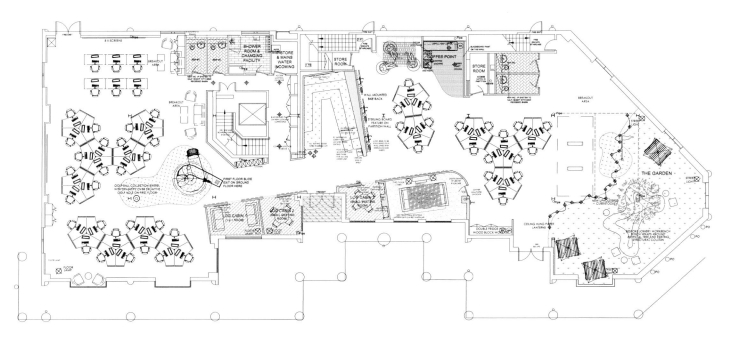

GROUND FLOOR

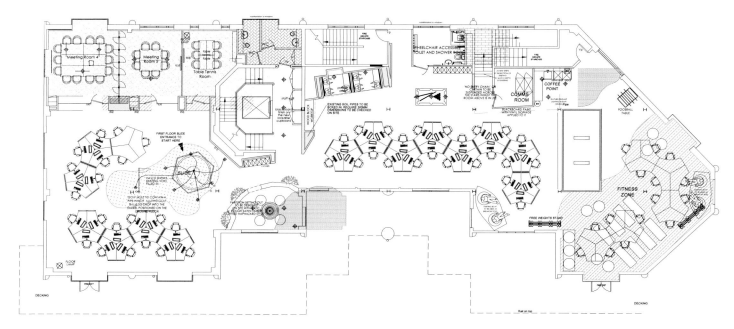

FIRST FLOOR

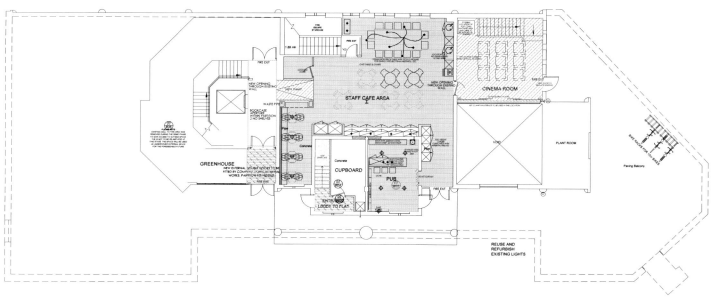

SECOND FLOOR

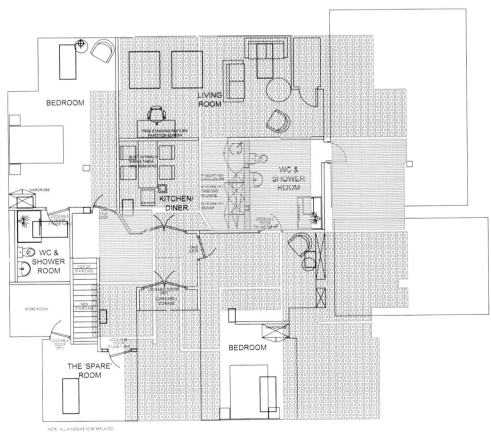

THIRD FLOOR

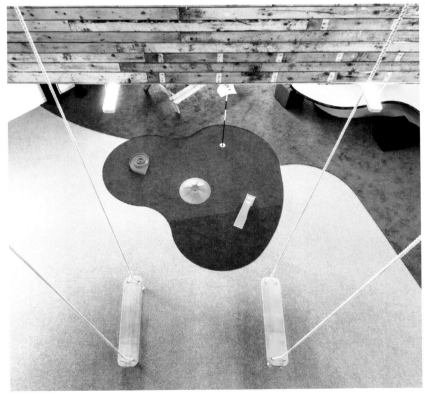

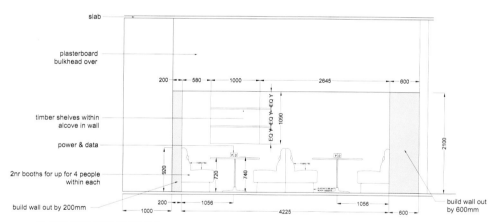

C - PROJECT AREA SEATING ELEVATION
Scale 1:20 @ A1

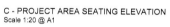
LOCATION PLAN N.T.S

LOCATION PLAN N.T.S

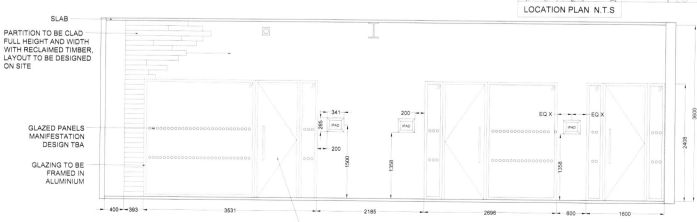

F - MEETING ROOM ELEVATION
Scale 1:20 @ A1

MEETING ROOM DOORS TO BE CONSTRUCTED WITH RECYCLED SCAFFOLDING BOARDS

INDEX

D+Z ARCHITECTEN+ PROJECTMANAGERS

D+Z Architecten+Projectmanagers is a privately owned independent project management and interior design company, founded in 1994 and employing 31 project specialists. D+Z offers turnkey solutions, guaranteeing delivery within confines of planning and budget. They take full responsibility for all parts of a project from the initial preparation phases to the very last details when completing the project and they allow their principals maximum insights and monitoring capabilities throughout the project.

D+Z has an in house architect studio, employing 10 (interior) architects. Traditional or innovative, functional or representative, based upon our principals DNA D+Z has created a great variety of office concepts and interior designs. The designs often have a distinctive character according the famous 'Dutch Design'.

D+Z is proud to have successfully completed complex relocation projects for international companies such as: Apple Benelux, Nike, McAfee International, VimpelCom, Cargill and Kia Motors.

137KILO

137kilo is a studio that operates in many scales from object and interiors to architecture. Their design process is based on strong concepts, rigorous analysis and contextual storytelling.

They have been working for private and public clients, both in Poland and abroad since 2006.

SPACE & SOLUTIONS

UK based Design & Build, Office and Relocation Company Space & Solutions has been designing & building offices that people love to work in since 2008. Specializing in cutting edge, modern and innovative office interiors, their highly experienced and personable team work closely with clients nationally and internationally to transform their business and aspirational objectives into inspiring and functional work environments. Not only does the team look to maximize workspace & efficiency, they also design workspaces that are comfortable and easy to work in which in turn has a positive impact on staff happiness and overall productivity of the organization.

CLIVE WILKINSON ARCHITECTS

Clive Wilkinson Architects is a distinguished architecture and design practice that collaborates with progressive clients to envision and design new buildings and environments that support, enhance and reinvigorate contemporary life.

Through a humanistically centered work process, the firm strives to connect people, shape relationships and empower organizations to produce new invigorating forms of human community. The firm has acquired a depth of experience for the enrichment of future projects by working with many of the world's most creative companies and institutions over the course of twenty years.

EDG

edg is devoted to enhancing the value between human and spatial environments and to be a first-class enterprise bringing new standards to the industry. Through superior design and construction quality services throughout the China Mainland, edg builds environmentally healthy, comfortable and over-fulfilling spaces for its clients.

edg is specialized in the interior design and construction of corporate offices, commercial spaces, boutiques, financial facilities and medical facilities. edg has the capability of bringing the most suitable design and most current building methodologies and technologies to each project based on its rich experience in international design and construction.

KINZO

Founded in 1998 as a label for film, design, architecture, brand and event projects, KINZO has become internationally established for its unmistakable style and grown to be one of the ambassadors of Berlin design avantgarde. Starting from interior projects for shops, lofts and offices, fairs and exhibitions as well as design for events and film sets, the three architects and founders of the company Karim El-Ishmawi, Chris Middleton and Martin Jacobs have become famous for their distinctive, sport and science-fiction inspired elegance. In 2008, KINZO successfully ventured into product design: Its office furniture programme KINZO AIR was awarded the red dot design award and nominated for the 2009 German State Design Price.

ZA BOR ARCHITECTS

za bor architects is a Moscow based architectural office founded in 2003 by principals Arseniy Borisenko and Peter Zaytsev.

The bureau's objects are created mainly in contemporary aesthetic. What distinguishes them is an abundance of architectural methods used both in the architectural and interior design, as well as a complex dynamical shape which is a visiting card of za bor projects. The interiors demonstrate this feature especially brightly, since for all their objects architects create built-in and free standing furniture themselves.

za bor architects have been involved in more than 60 projects including residential houses, an office building, a cottage settlement, many offices. Among the clients of za bor architects there are media, business and government companies such as Forward Media Group, Yandex, Inter RAO UES, Moscow Chamber of Commerce and Industry and others.

JUMP STUDIOS

Jump Studios is a London based architecture and design practice. Established in 2001 by Shaun Fernandes and Simon Jordan, Jump Studios has completed several award-winning projects to date for clients including Nike, Levi's, Innocent Smoothies, Red Bull, Adidas, Wieden + Kennedy, The Science Museum and Honda. Jump is currently working on projects for Starwood Hotels and The Marketing Store among others.

O.S.O ARCHITECTURE & INTERIOR DESIGN

O.S.O Architecture & Interior Design is an Istanbul-based architectural studio established by Okan Bayık, Serhan Bayık and Ozan Bayık in 2007. A strong emphasis is given to the critical design process within the studio; They resist predetermining architectural solutions to a client's brief prior to a thorough investigation of each project's unique situation.

Their criteria for design is to pay close attention to contemporary design methods, new materials and the economic considerations of the client. They believe that a good project must combine all these areas. Moreover, the most important thing for them is to look at a project from a new perspective. O.S.O Architecture specializes in architecture, interior design and project management. Their aim is to create examples which are not only unique, but also combining efficiency, economy design.

M MOSER ASSOCIATES

M Moser Associates is an interiors and architecture firm specializing in workplace strategy and design. Their Integrated Project Delivery approach optimizes diverse project team capabilities to deliver solutions that effectively facilitate the way clients work. Since 1981, M Moser has completed more than 4,500 workplace projects, ranging from interiors to corporate base buildings, campuses and tech-intensive facilities like labs and data centers. Their 700+ staff includes planners, interior designers, architects, engineers, construction professionals and specialists in IT integration and sustainable design, all working collaboratively from 17 global locations.

KLEIN DYTHAM ARCHITECTURE

Klein Dytham architecture, established by Royal College of Art graduates Mark Dytham and Astrid Klein is based in Tokyo. The Japanese thirst for the new, sensitivity to material and detail in crafting, and an ever-changing urban environment nourish KDa's ideas and production. Their client list includes Google, Tsutaya, Sony, Nike, and Shiseido. Klein Dytham co-founded SuperDeluxe, a art space which is the birth place of PechaKucha Night, a show-and-tell evening where each presentation is 20 slides x 20 seconds. PechaKucha Nights taking place in over 630 cities around the world in 2013.

COAST OFFICE

COAST Office is a multidisciplinary design practice active internationally in the fields of architecture, interiors and exhibitions. Founded and directed by Zlatko Antolovic and Alexander Wendlik, the core work of COAST Office is to develop architecture and spaces that engage all the senses and foster interaction between humans, space, nature and technology – bringing together past, present and future. The diversity of their work is the result of their conceptual approach that explores and analyzes history, cultural and traditional backgrounds, nature, society, technologies and brand identities to find a specific solution to each client, task and location.

PENSON

PENSON focuses on providing talented architecture, interior design, structural, civil, mechanical & electrical engineering consultancy services, specializing in all sectors of buildings & uses.

They are unique because their chartered architects, designers & engineers sit together, to simplify communication whilst improving coordination & efficiency. They grow by working hard to create award winning solutions, whilst paying close attention to every project's commercial & deliverable needs.

Their team unites London's best young talent leading the field of design & sustainability. They help to judge international design competitions, speak at conferences, teach architecture & write articles for some of the world's leading design journals. PENSON also host frequent events for numerous charities for fun & enhanced team building.

LORCAN O'HERLIHY ARCHITECTS

Lorcan O'Herlihy Architects (LOHA) is committed to engaging the complexities of contemporary society.

LOHA approaches their work with ruthless optimism. They believe that pragmatics can inspire a creativity that is resourceful, authentic, and profound. They steep their process in rigorous exploration and then they improvise; They pair poetry with reason to achieve essential solutions.

LOHA builds their ideas within real world politics. They work with the conviction that architecture dynamically integrates and can incrementally enhance lives.

Since 1990, LOHA has built over 75 projects in three continents ranging in typology from institutional buildings to bus shelters, and from large-scale developments to single-family homes. LOHA has garnered over a hundred awards, including the 2010 AIA Los Angeles Firm of the Year.

camenzindevolution
CAMENZIND EVOLUTION

Winner of the RIBA Worldwide Award, the Young Architect of the Year Award and the International Design Award, the Swiss architectural studio Camenzind Evolution has placed its selves firmly within the new generation of up-and-coming architects. Fundamental to their very diverse work is the fusion of emotional design ingenuity with Swiss quality. Having gained many years of international experience in the architectural offices of Nicholas Grimshaw in London and Renzo Piano in Paris, in 2004 Stefan Camenzind founded the architectural studio Camenzind Evolution which has evolved into a partnership with Russian born architect Tanya Ruegg-Basheva. The studio has grown to 15-20 architects today.

Specialized Fields

Camenzind Evolution are the designers of the extraordinary workplace designs of the Google Offices in Dublin, Tel Aviv, Zurich, Moscow, Oslo & Stockholm, and have also created and designed outstandingly unique and innovative workplace solutions for Unilever, Credit Suisse and other renowned companies. They celebrate an user-inclusive design approach, using psychological research methods to create a work environment which suits people's personalities, matches their senses, energizes their bodies and minds – and ultimately, stimulates and inspires people to their highest levels of creativity and innovation.

SID LEE ARCHITECTURE

Sid Lee Architecture is a partnership between Sid Lee, a global commercial creativity company, and seasoned architects Jean Pelland and Martin Leblanc.

Their core belief is that architecture can shape the identity of companies, organizations and communities. Their novel approach consists of leveraging the power of multidisciplinary thinking by teaming architects with artisans from different fields to give rise to a completely new vision of architecture as a vehicle to create immersive experiences that go beyond form and function by integrating rich narratives.

Put simply, they mix architecture and branding. Sid Lee Architecture's uniqueness is rooted in the understanding of this potent interrelation.

JOHNSON CHOU INC.

Johnson Chou Inc. is an internationally recognized interdisciplinary design practice encompassing architecture and interiors, furniture, industrial and graphic design – a body of work characterized by conceptual explorations of narrative, transformation and multiplicity. By creating "narratives of inhabitation", be it a residence, office or retail space, the studio's projects are characterized by forms imbued with metaphoric content and richness of detail – objects and spaces that are "portraits" of their clients.

MEGABUDKA

Megabudka is a wonderful office/workshop/creative association/team/group of architects from Moscow. Company was founded in 2008 by graduates of the Moscow Architectural Institute. For 4,5 years megabudka has won (1st, 2nd, 3rd places) more than 10 architecture competitions and got 5 honorable mantion.

They're trying to prove to the world that in Russia there is a qualitative ideological architecture.

blitz DESIGN BLITZ

Design Blitz was founded in 2009 during one of the worst recessions in history. The founders of Design Blitz, Seth Hanley and Melissa Wallin, saw the economic decline as an opportunity to reinvent the practice of architecture using the business and project implementation philosophies of their successful peers in the technology sector. Eschewing "business-as-usual" for a leaner, agile work-flow, Design Blitz flourished while other companies shuttered their doors. Design Blitz has since grown into a dynamic group of creative architects and designers with eclectic backgrounds who are obsessed with creating really cool flexible environments that are fresh and innovative. As a full service architecture and interior design firm, they provide the complete range of architectural services required to take a project from programming through construction. Though their focus is the built environment, Design Blitz is committed to total design solutions – balancing buildings, branding, and experiences.

AEI

Arquitectura e Interiores (Aei) is a full service Colombian architectural company, specializing in award-winning corporate and commercial interior fitout. They are a founding member of a Latin American partnership of industry-leading design & construction firms; together they maintain a staff of over 1,000 architects in 11 countries throughout the region and offer a full range of services that include design, construction, project management, change management, as well as LEED consultation.

They create office spaces that are not only aligned to the core business of their clients, but also respond to the changing dynamics of the workplace. They offer an integrated, coordinated and flexible service that exceeds their clients' expectations.

Via Architettura
V ARC

V Arc is an innovative and responsive design practice, focused on collaboration between staff and client. Frank Bambino and Sergio Pettrica established V Arc more than 14 years ago with the aim to produce architecture and interiors that make a significant contribution to the urban environment.

V Arc's portfolio includes projects across Australia in Corporate, Retail & Hospitality, Aged & Health Care, Residential & Resort, Institutional. This includes the recently completed global headquarters for Telstra, a 40 floor refurbishment for 50,000 staff as part of Australia's largest leasing negotiation in history.

V Arc have a 30 strong team with offices in Melbourne and Sydney, and associated offices in Brisbane, Gold Coast, Adelaide and Perth. V Arc are driven by the desire to operate around a studio philosophy, working as a team with Their clients to be first choice and to achieve the best results and producing architecture and interiors that make a significant impact on the end users.

LEMAY

Established in 1957 and boasting over 55 years of success, Lemay is one of the largest integrated design firms in Canada, with offices in Montréal, Québec and Toronto, as well as in Algeria and Costa Rica. Backed by its team of multidisciplinary designers, the firm offers a vast range of creative services in architecture, urban design, interior design and graphic design.

Lemay distinguishes itself from its competitors with distinctive, original and memorable projects of all scales and complexities. The fruit of collective knowledge supported by a half-century of experience, these projects are also the result of the firm's unique creative approach. This driving force is animated by a constant search for excellence and a firm volition to innovate and question conventional practices, particularly those pertaining to the environment. A true partner to its clients, Lemay offers a memorable service experience founded on a synergy of expertise, teamwork and the talent of its multidisciplinary creators.

INNOCAD

INNOCAD carries out numerous projects in housing, offices, health service, retail, interior and product design, e.g. Flur20 a social housing project, Rolling Stones – Landeszeughaus Graz, Samsung Vienna headquarters, Microsoft Vienna headquarters awarded with the "great place to work 2012", best architects label 13 and with the 34th annual interior award, XAL Corporate Architecture, including the XAL Competence Center awarded with the 34th annual interior award and trade fair booths and showrooms around the globe, penthouse in New York, neurological out-patient department at LKH Feldbach awarded with the AIT award and the GerambRose, residential and office building "Am Kai" Graz, residential and office building "Rose am Lend", nominated, among other things, for best architects label 2013 and for the Styrian Provincial Construction Prize, Voestalpine headquarters, and the office and residential building Golden Nugget. The latter received the 2005 TECU Architecture Award, Special Prize for Residential Buildings, and the 2006 International VELUX Award for architecture.

ERGİNOĞLU & ÇALIŞLAR ARCHITECTS

Erginoğlu & Çalışlar Architects is an Istanbul based, independent firm of architects founded in 1993 by Hasan Çalışlar and Kerem Erginoğlu. The firm specializes in urban planning, architecture, and interior design projects together with providing assistance for planning applications.

The ethos of the company is to view each project within its individual context and contribute to it through innovative architectural solutions. The firm has vast experience gained through successful completion of a wide variety of both national and international projects on a range of scales.

STUDIO SARAH WILLMER ARCHITECTURE

Studio Sarah Willmer Architecture is an award-winning nationally and internationally published architecture and interiors firm based in San Francisco. The primary premise of the studio is a strong belief that collaboration with the client creates a more dynamic and unique design solution for each project. The principles of simplicity, of spaciousness and light, of synthesis and function are realized in each project. Three principals guide the design process with each client: the building's structure as a primary expression of space, the control and infusion of natural light, and an environmentally sensitive and elegant material palette. The creative energy at Studio Sarah Willmer is vitalized by the extensive experience of its senior staff and youthful enthusiasm of recent grads from University of California, Berkeley and California College of the Arts. The team is tight and selected for their unending commitment to quality architecture and critical design.

IWAMOTOSCOTT ARCHITECTURE

IwamotoScott Architecture is a San Francisco based practice led by Lisa Iwamoto and Craig Scott. IwamotoScott's design process proceeds from the belief that each project can achieve a unique design synthesis. Dedicated to intensive design research, the practice engages in projects at multiple scales and in a variety of contexts consisting of full-scale fabrications, museum installations and exhibitions, theoretical proposals, competitions and commissioned design and building projects. Conceptual themes of the work focus on strategies of adaptation, and intensifying the experiential and performance based qualities of architecture. These conceptual directions are pursued through exploration in computational techniques, hands-on prototyping, investigation into material behavior, as well as empirical studies of the experiential effects afforded by formal, spatial and material strategies. IwamotoScott are committed to employing new technologies in the production of architecture, capitalizing on their potential to inform the design process and gear the evolution of the designs.

ACKNOWLEDGEMENTS

We would like to thank everyone involved in the production of this book, especially all the artists, designers, architects and photographers for their kind permission to publish their works. We are also very grateful to many other people whose names do not appear on the credits but who provided assistance and support. We highly appreciate the contribution of images, ideas, and concepts and thank them for allowing their creativity to be shared with readers around the world.